South University Library
Richmond Campus
2151 Old Brick Road
Glen Allen, Va 23060

MAR 0 6 2018

UNIVERSITIES AS COMPLEX ENTERPRISES

STEVENS INSTITUTE SERIES ON COMPLEX SYSTEMS AND ENTERPRISES

Title: *Modeling and Visualization of Complex Systems and Enterprises: Explorations of Physical, Human, Economic, and Social Phenomena*
Author: William B. Rouse
ISBN: 9781118954133

Title: *Perspectives on Complex Global Challenges: Education, Energy, Healthcare, Security, and Resilience*
Editors: Elisabeth Paté-Cornell and William B. Rouse with Charles M. Vest
ISBN: 9781118984093

UNIVERSITIES AS COMPLEX ENTERPRISES

How Academia Works,
Why It Works These Ways,
and Where the University Enterprise
Is Headed

WILLIAM B. ROUSE

WILEY

Copyright © 2016 by John Wiley & Sons, Inc. All rights reserved

Published by John Wiley & Sons, Inc., Hoboken, New Jersey
Published simultaneously in Canada

No part of this publication may be reproduced, stored in a retrieval system, or transmitted in any form or by any means, electronic, mechanical, photocopying, recording, scanning, or otherwise, except as permitted under Section 107 or 108 of the 1976 United States Copyright Act, without either the prior written permission of the Publisher, or authorization through payment of the appropriate per-copy fee to the Copyright Clearance Center, Inc., 222 Rosewood Drive, Danvers, MA 01923, (978) 750-8400, fax (978) 750-4470, or on the web at www.copyright.com. Requests to the Publisher for permission should be addressed to the Permissions Department, John Wiley & Sons, Inc., 111 River Street, Hoboken, NJ 07030, (201) 748-6011, fax (201) 748-6008, or online at http://www.wiley.com/go/permissions.

Limit of Liability/Disclaimer of Warranty: While the publisher and author have used their best efforts in preparing this book, they make no representations or warranties with respect to the accuracy or completeness of the contents of this book and specifically disclaim any implied warranties of merchantability or fitness for a particular purpose. No warranty may be created or extended by sales representatives or written sales materials. The advice and strategies contained herein may not be suitable for your situation. You should consult with a professional where appropriate. Neither the publisher nor author shall be liable for any loss of profit or any other commercial damages, including but not limited to special, incidental, consequential, or other damages.

For general information on our other products and services or for technical support, please contact our Customer Care Department within the United States at (800) 762-2974, outside the United States at (317) 572-3993 or fax (317) 572-4002.

Wiley also publishes its books in a variety of electronic formats. Some content that appears in print may not be available in electronic formats. For more information about Wiley products, visit our web site at www.wiley.com.

Library of Congress Cataloging-in-Publication data applied for

ISBN: 9781119244875

Printed in the United States of America

10 9 8 7 6 5 4 3 2 1

CONTENTS

Preface   xi

1   Introduction and Overview   1

   Overall Approach, 4
   Universities as Complex Systems, 5
   Complex Adaptive Systems, 6
   Universities as Complex Adaptive Systems, 8
      Nonlinear, Dynamic Behavior, 8
      Independent Agents, 8
      Goals and Behaviors That Differ or Conflict, 8
      Intelligent and Learning Agents, 9
      Self-Organization, 9
      No Single Point(s) of Control, 9
      Implications, 9
   Overview of Chapters, 10
      Chapter 1: Introduction and Overview, 10
      Chapter 2: Evolution of the Research University, 10
      Chapter 3: Mission and Structure, 10
      Chapter 4: Leadership and Governance, 11
      Chapter 5: Administration, 11
      Chapter 6: Money and Space, 11
      Chapter 7: Promotion and Tenure, 11
      Chapter 8: Education Programs, 12
      Chapter 9: Research and Intellectual Property, 12

Chapter 10: Rankings and Brand Value, 12
Chapter 11: Transformation Scenarios, 12
Chapter 12: Exploring the Future, 13
References, 13

2 Evolution of the Research University 15

Early Universities in Europe, 16
    University of Bologna, 16
    University of Paris, 17
    University of Oxford, 17
    University of Padua, 17
    University of Cambridge, 18
    Summary, 18
Early Universities in America, 19
Humboldt's Innovation, 19
Morrill Acts, 20
Bush and NSF, 22
Engineering Science, 24
Today's Research Universities, 24
Conclusions, 26
References, 27

3 Mission and Structure 29

Mission, 29
    Athletics, 31
Structure, 32
    Ecosystem: Society and Government, 32
    Structure: Campuses, Colleges, Schools, and Departments, 33
    Processes: Education, Research, and Service, 34
    Practices: Education, Research, and Service, 35
    Research Centers, 35
    Appointments, 38
Conclusions, 39
References, 39

4 Leadership and Governance 41

Leadership, 42
    Leadership and Change, 42
    Leadership and Time, 44
    Stewards of the Status Quo, 46
    Leading Research Centers, 48
    Leadership Experiences, 49
Governance, 51

CONTENTS vii

        Governing Boards, 51
        Administration and Faculty, 51
        Other Players, 52
        Governance Experiences, 52
Conclusions, 55
References, 55

5 Administration   57

Number of Administrators and Costs, 59
Performance Evaluation, 61
Conflict Management, 64
        Conflicts within Organizations, 64
        Conflicts across Organizations, 64
        People Conflicts, 65
Compliance and Abuse, 66
Marketing and Communications, 66
        Book Series, 67
The Costs of Conformity, 67
Conclusions, 68
References, 69

6 Money and Space   71

Economics of Higher Education, 72
        Value of Education, 72
        Economists' Views, 73
        Government Subsidies, 74
        Higher Education Bubble, 74
        Public Endowment, 75
Costs of Higher Education, 75
        Cost Disease, 75
        Cost Analyses, 76
        Indirect Costs, 77
        Staffing Patterns, 77
        Student and Institutional Debt, 78
Revenue: Tuition, 79
Revenue: Government Dependencies, 80
Revenue: Fundraising, 81
        Fundraising Experiences, 82
        Summary, 83
Lessons Learned, 83
Overall Economic Model, 85
Space, 86
Conclusions, 87
References, 87

7 Promotion and Tenure 89

Nature and Roles of Faculty, 90
    Academic Disciplines, 90
    Faculty Impact, 91
    TT versus NTT Faculty, 92
    Availability of Faculty Positions, 92
    Faculty Turnover, 93
Nature of Tenure Decisions, 95
Promotion and Tenure Experiences, 97
    What Really Counts, 98
    Making the Case, 98
Model of Tenure Decision Making, 99
Conclusions, 103
References, 103

8 Education Programs 105

Stem Challenges, 106
Student Population, 106
Value of Education, 107
Degree Programs, 108
Curricula and Courses, 109
Delivery of Education, 110
Teaching Experiences, 112
Workforce Model, 114
Conclusions, 118
References, 119

9 Research and Intellectual Property 121

Challenges, 122
    Peer Review, 122
    Bibliometrics, 122
    Funding, 124
Research Experiences, 125
    Libraries and Networks, 125
    Limits of Modeling, 125
    Healthcare Delivery, 126
    Interactive Visualization, 126
    Government Sponsors, 127
    Industry Sponsors, 127
Research Model, 128
    Submission of Articles, 128
    Citation of Articles, 130
    Submission of Proposals, 131
    Overall Model, 132

CONTENTS ix

   Intellectual Property, 135
     Spin-Off Experiences, 135
   Conclusions, 137
   References, 137

10 Rankings and Brand Value   **139**

   Ranking Schemes, 140
   Example of Moving Up, 142
   Determinants of Rankings, 143
   Brand Value, 146
     Model of Brand Value, 146
     More on Metrics, 148
     Example, 149
   Conclusions, 150
   References, 151

11 Transformation Scenarios   **153**

   Forces for Change, 153
     Costs and Benefits, 154
     Globalization, 154
     A Tsunami of Talent, 156
     Technology, 157
   Organizational Change, 158
     Theory and Practice, 159
   Four Scenarios, 160
     Driving Forces, 161
     Clash of Titans, 161
     Hot, Flat, and Crowded, 162
     Lifespan Mecca, 162
     Network U., 163
     Implications, 163
   Transforming Academia, 164
     Clash of Titans, 165
     Hot, Flat, and Crowded, 165
     Lifespan Mecca, 166
     Network U., 166
   How Change Happens, 167
   Conclusions, 168
   References, 169

12 Exploring the Future   **173**

   Sensitivity Analyses, 176
   Scenario Variations, 180
     Clash of Titans, 180

Hot, Flat, and Crowded, 180
 Lifespan Mecca, 181
 Network U., 181
 Projections, 182
Policy Implications, 185
 Across Scenarios, 185
 Within Scenarios, 186
 Summary, 187
Extensions, 187
Conclusions, 189
References, 189

Index 191

PREFACE

I first entered academia in 1965, over 50 years ago, as a freshman at the University of Rhode Island. Since then, I have served on the faculties of five universities in the United States and Europe, participated on advisory boards of numerous universities, been a consultant to several universities and, overall, been involved with over 50 universities globally. This has resulted in a wealth of fascinating, enlightening, and sometimes frustrating experiences. The intellectual community is often captivating, while the rigidity and resistance to change are sometimes daunting. This book tells the story of those experiences, woven into more expository material on a thousand years of academic organizations. This book also provides the basis for projecting where this fascinating and frustrating enterprise is headed.

A few years ago, I led a graduate course on "transforming academia." The class included eight PhD students and four faculty members. We explored the historical roots of universities and contemporary challenges for academia. The students led this exploration. I organized the syllabus and compiled resource materials, but the PhD students each gave two lectures, all of which were followed by a usually intense discussion. The course culminated with the students presenting to the president and provost their recommendations for how Georgia Tech should strategically think about the future. This presentation was scheduled for 1 hour, but the meeting stretched to 2 hours due to the many questions and much discussion. The findings of this semester-long exploration are laced throughout this book.

I have been involved in many strategic planning activities at numerous universities, either as a faculty member or as a paid consultant. In my most recent engagement, several senior administrators asked, "What shall we assume about the future?" Discussing this question quickly led to the conclusion that any specific prediction would likely be wrong. Consequently, I developed four scenarios for the future of academia that are discussed in this book. The faculty embraced these scenarios as good portrayals of alternative futures. Subsequently, all the planning teams were asked to assess the merits of their plans relative to each of the four scenarios. They struggled with this assignment and invariably ended up focusing on what could be called "business as usual on steroids." When asked why they ignored the other three scenarios, they responded that they did not know how to think about such different views of academia. They were stuck.

Academic enterprises do seem to be stuck. Increases in the costs of higher education have completely outstripped inflation. Academia has become the poster child for runaway costs, replacing healthcare, which now seems more or less under control. This book explores the nature of academic enterprises, including why they work the way they do and where such enterprises are headed. The goal, however, is not prediction. Instead, the objective is insights into where change can and will happen.

Many people have helped me in my exploration of academia over many decades. My experience at MIT was and continues to be mentored by Dick Larson, Joel Moses, and Tom Sheridan and the late Chuck Vest. My brief experience at Tufts was benefited by the expertise of Percy Hill. My mentors at Illinois were B.T. Chao, Bob Chien, Dan Drucker, and Helmut Korst. My one year in Delft changed my life—I am indebted to Henk Stassen and more recently Theo Toonen. Two stops at Georgia Tech provided many mentors—Jean-Lou Chameau, Wayne Clough, Steve Cross, Rich DeMillo, Don Giddens, Paul Griffin, John Jarvis, Leon McGinnis, Mike Thomas, and John White. Mentors at Stevens have included Tony Barrese, Michael Bruno, Ralph Giffin, Nariman Farvardin, George Korfiatis, and Dinesh Verma. Other colleagues related to singular experiences rather than long-term appointments. These included Berkeley (Lee Schruben), Carnegie Mellon (Duane Adams), Case Western Reserve (Simon Ostrach), George Mason (Andy Sage), Georgetown (Spiros Dimolitsas), Kassel (Gunnar Johannsen), North Carolina State (Paul Cohen and Ed Fitts), Penn State (Paul Griffin), Stanford (Elisabeth Pate-Cornell), Tsinghua (Gavriel Salvendy), Tokyo Institute of Technology (Kinji Mori), and the University of California at San Diego (Hal Sorenson). Opportunities are required to create experiences, and I am indebted to these many colleagues for providing me such a rich set of experiences.

WILLIAM B. ROUSE
WEEHAWKEN, NJ
NOVEMBER 2015

1

INTRODUCTION AND OVERVIEW

Higher education has become the "poster child" for out-of-control costs, replacing healthcare, which now seems more or less under control. Tuition increases have far outpaced increases of the overall cost of living. This is due to the relative decline of public support for higher education, while administrative costs have been steadily growing much faster than the costs of teaching and research. A primary enabler and consequence of this cost growth has been student debt levels that exceed the total credit card debt in the United States.

We need to get a grip on the economics of higher education, with a goal of transforming the system to improve the overall value proposition. Academia provides a wide variety of offerings that serve a diverse mix of constituencies. Delivery processes for these offerings can be quite creative but are often burdened with inefficiencies. This is complicated by academic governance processes, which can be overwhelming, more so when the university is also a public sector agency. Yet, universities are basically well-intentioned, creative, and committed. There is much to build on but nevertheless much to overcome.

Figure 1.1 depicts a multilevel architecture of academic enterprises (Rouse, 2015). The practices of education, research, and service occur in the context of processes, structure, and ecosystem. Understanding the relationships among practices, processes, structure, and ecosystem provides the basis for transforming academia, leveraging its strengths, and overcoming its limitations. In this book, I explicitly address these relationships in terms of both conceptual and computational models of academic enterprises.

Universities as Complex Enterprises: How Academia Works, Why It Works These Ways, and Where the University Enterprise Is Headed, First Edition. William B. Rouse.
© 2016 John Wiley & Sons, Inc. Published 2016 by John Wiley & Sons, Inc.

The architecture in Figure 1.1 helps us to understand how various elements of the enterprise system either enable or hinder other elements of the system, all of which are embedded in a complex behavioral and social ecosystem. Practices are much more efficient and effective when enabled by well-articulated and supported processes for delivering capabilities and associated information, as well as capturing and disseminating outcomes.

Processes exist to the extent that organizations (i.e., campuses, colleges, schools, and departments) invest in them. These investments are influenced by economic models and incentive structures and are made in pursuit of competitive positions and economic returns. These forces hopefully coalesce to create an educated and productive population, at an acceptable cost.

When we employ Figure 1.1 to understand relationships among universities, the interesting phenomenon in Figure 1.2 emerges. The hierarchical structure of Figure 1.1 dovetails with the heterarchical nature of academic disciplines. The dotted rectangle in Figure 1.2 represents how faculty disciplines both compete and define standards across universities.

The disciplines define the agenda for "normal" science and technology, including valued sponsors of this agenda and valued outlets for research results. Members of faculty disciplines at other universities have an enormous impact on promotion and tenure processes at any particular university. Such professional affiliations also affect

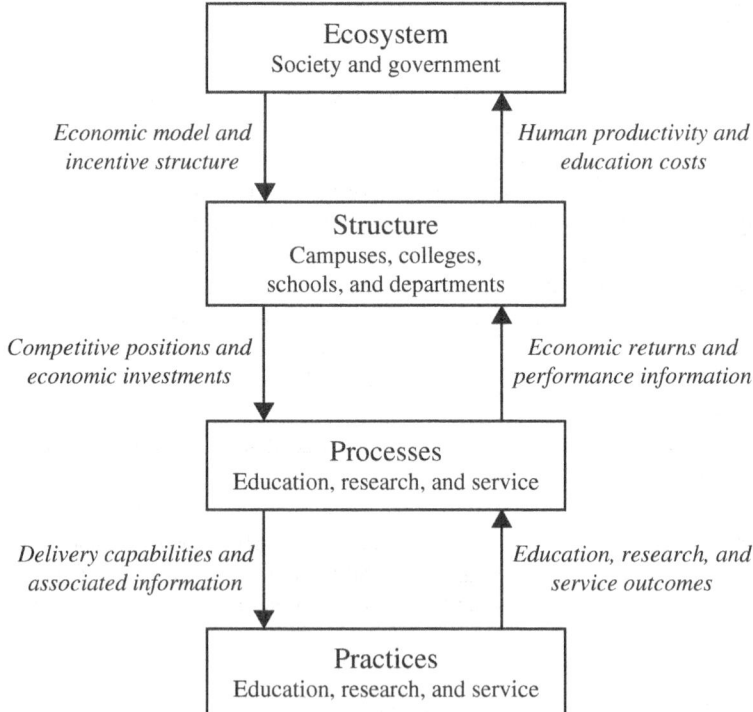

FIGURE 1.1 Multilevel architecture of academic enterprises.

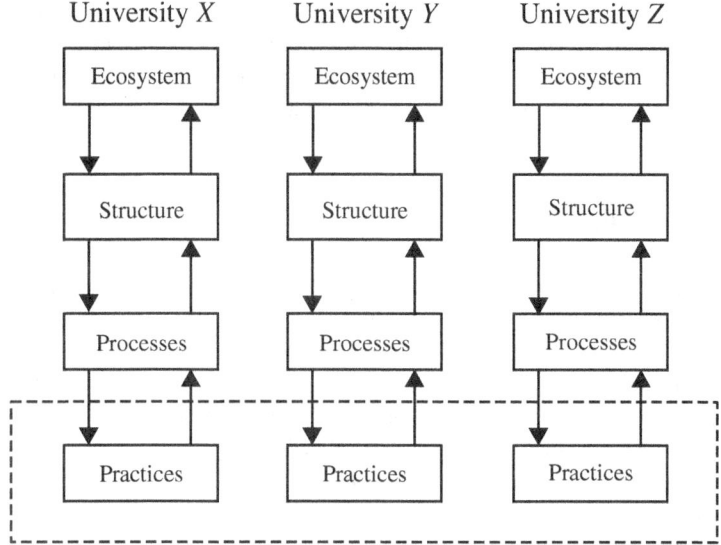

FIGURE 1.2 Hybrid multilevel architecture of academia.

other types of enterprises (e.g., healthcare). However, universities seem to be the only enterprise that allows external parties to largely determine who gets promoted and tenured internally. This has substantial impacts on understanding and modeling the performance of any particular university.

More specifically, the standards set at the discipline level determine:

- Agenda for "normal" science and technology
- Valued sponsors of this agenda
- Valued outlets for research results

Consequently, almost everyone chases the same sponsors and journals, leading to decreasing probabilities of success with either. In addition, each faculty member produces another faculty member every year or so, swelling the ranks of the competitors. Recently, retirements are being delayed to refill individuals' retirement coffers, which decrease numbers of open faculty slots.

As probabilities of success decrease, faculty members write an increasing number of proposals and submit an increasing number of journal articles, resulting in constantly increasing costs of success and congested pipelines, which foster constantly increasing times until success. Bottom line is less success, greater costs, and longer delays. The models discussed in this book enable exploration of these phenomena.

Universities can hold off these consequences by hiring fewer tenure-track faculty members, that is, using teaching faculty and adjuncts. But this will retard their march up the rankings and hence slow the acquisition of talented students, who will succeed in life and later reward the institution with gifts and endowments. The trade-off between controlling cost and enhancing brand value is explored in this book.

Alternatively, universities can pursue "niche dominance" and only hire tenure-track faculty in areas where they can leapfrog to excellence. This will, unfortunately, result in two classes of faculty—those on the fast track to excellence and those destined to teach a lot. The first class will be paid a lot more because of the great risks of their being attracted away to enhance other universities' brands.

OVERALL APPROACH

Many contemporary commentators have made similar observations to those offered in this introduction, although with less emphasis on the structure and processes of the enterprise. For example, Lombardi (2013) provides an exposition of the basic blocking and tackling of academia, primarily from a more general perspective than science and technology. This is relevant in terms of the nature of academic guilds portrayed and the multilevel nature of governance. He also provides a useful discussion of performance and quality management. However, his book does not articulate an overall enterprise perspective.

Christenson and Eyring (2011) explore how universities can find innovative, less costly ways of performing their uniquely valuable functions and thereby save themselves from decline. The authors outline the history of Harvard University and how various aspects of academia were defined by Harvard leadership in response to issues and opportunities of the times. They explore the strategic choices and alternative ways in which traditional universities can change to ensure their ongoing economic vitality. They emphasize the need for universities to address key trade-offs and make essential choices as they decide how to compete.

DeMillo (2011) addresses the challenges faced by "the middle," the 2000 universities that are not part of the "elite"—those with $1 billion plus endowments—and also face stiff competition from for-profit online universities. Computer-based and online technologies, such as MOOCs, and new student-centric business models are discussed. The book culminates in 10 rules for twenty-first-century universities, expressed in terms of defining value and becoming an architect of how this value is delivered. Some of the ideas presented are very relevant to the issues addressed in this book.

These three books are representative of a vast literature on higher education. Many other sources are referenced throughout this book as specific issues are discussed. This book leverages and integrates this material in the following ways:

- The university is viewed as an enterprise system competing in a complex economic, political, and social environment where both competition and collaboration are prevalent
- Examples are drawn from a broad range of experiences with over 50 university enterprises in the United States, Europe, Asia, Africa, and Latin America
- A multilevel architectural view of universities as enterprises enables identifying interactions and opportunities across levels of this architecture, including both avenues and barriers to fundamental change

I address university enterprises both qualitatively and quantitatively. Qualitative expositions draw from history, public policy, economics, etc. Many, but not all, of these expositions set the stage for quantitative models of the phenomena of interest. The quantitative models are valuable for addressing various "what-if" questions discussed in this book.

Examples of such models include use of the enterprise architecture in Figures 1.1 and 1.2 for computational modeling of enterprise performance and economic models of key trade-offs such as the mix of tenure-track versus nontenure-track faculty members. Discounted cash flow models are employed to reflect the time value of money involved in such decades-long investments.

Another example is statistical models of the dynamics of the *US News & World Report* ranking system, in particular time series models of the lags in ranking changes following university investments in faculty, facilities, and other strategic investments. There is a frequently expanding range of university ranking systems, for example, *Financial Times* World University Rankings and Shanghai Jiao Tong University Academic Ranking of World Universities. The various ranking schemes are contrasted in terms of purpose and attributes.

UNIVERSITIES AS COMPLEX SYSTEMS

Universities certainly seem complex—so many stakeholders, agendas, and often conflicting priorities. What type of complex system is a university or, more broadly, universities collectively? They certainly are not similar to airplanes, factories, and process plants, all of which were engineered to achieve specified objectives. They are a subset of the society in general, but they do not seem as complex as the overall society.

Snowden and Boone's (2007) Cynefin Framework includes simple, complicated, complex, and chaotic systems. Simple systems can be addressed with best practices. Complicated systems are the realm of experts. Complex systems represent the domain of emergence, as discussed in the following text. Finally, chaotic systems require rapid responses to stabilize potential negative consequences.

The key distinction with regard to the discussions in this book is complex versus complicated systems—simple certainly is not warranted and we hope to avoid chaotic. There is a tendency, Snowden and Boone contend, for experts in complicated systems to perceive that their expertise, methods, and tools are much more applicable to complex systems than is generally reasonable.

Poli (2013) also elaborates the distinctions between complicated and complex systems. Complicated systems can be structurally decomposed. Relationships can be identified, either by decomposition or in some cases via blueprints. "Complicated systems can be, at least in principle, fully understood and modeled." Complex systems, in contrast, cannot be completely understood or definitively modeled. He argues that biology and all the human and social sciences address complex systems.

Poli also notes that problems in complicated systems can, in principle, be solved. The blueprints, or equivalent, allow one to troubleshoot problems in complicated systems. In contrast, problems in complex systems cannot be solved in the same way. Instead, problems can be influenced so that unacceptable situations are at least partially ameliorated.

These distinctions are well taken. Complicated systems have often been designed or engineered. There are plans and blueprints. There may be many humans in these systems, but they are typically playing prescribed roles. In contrast, complex systems, as they define them, typically emerge from years of practice and precedent. There are no plans and blueprints. Indeed, much research is often focused on figuring out how such systems work. Good examples are human biology and large cities.

COMPLEX ADAPTIVE SYSTEMS

The nature of human and social phenomena within complex systems is a central consideration. Systems where such phenomena play substantial roles are often considered to belong to a class of systems termed complex adaptive systems. This class of systems includes healthcare delivery (Rouse, 2000, 2008), urban systems (Rouse, 2015), and, as discussed in the following text, universities

Complex adaptive systems have the following characteristics:

- They tend to be **nonlinear, dynamic** and do not inherently reach fixed equilibrium points. The resulting system behaviors may appear to be random or chaotic
- They are composed of **independent agents** whose behaviors can be described as based on physical, psychological, or social rules rather than being completely dictated by the physical dynamics of the system
- Agents' needs or desires, reflected in their rules, are not homogeneous and, therefore, their **goals and behaviors are likely to differ or even conflict**—these conflicts or competitions tend to lead agents to adapt to each other's behaviors
- Agents are **intelligent and learn** as they experiment and gain experience, perhaps via "meta" rules, and consequently change behaviors. Thus, overall system properties inherently change over time
- Adaptation and learning tend to result in **self-organization** and patterns of behavior that emerge rather than having been designed into the system. The nature of such emergent behaviors may range from valuable innovations to unfortunate accidents
- There is **no single point(s) of control**—system behaviors are often unpredictable and uncontrollable, and no one is "in charge." Consequently, the behaviors of complex adaptive systems usually can be influenced more than they can be controlled

As summarized in Table 1.1, understanding and influencing systems having these characteristics creates significant complications. In general, leaders, who have more

TABLE 1.1 Implications of Complex Adaptive Systems

	Traditional System	Complex Adaptive System
Roles	Management	Leadership
Methods	Command and control	Incentives and inhibitions
Measurement	Activities	Outcomes
Focus	Efficiency	Agility
Relationships	Contractual	Personal commitments
Network	Hierarchy	Heterarchy
Design	Organizational design	Self-organization

influence than power, replace managers. These leaders craft incentives and inhibitions to influence behaviors—they cannot employ command and control. The goal is to influence outcomes because activities cannot be controlled and are often unobservable.

The focus is on organizational agility rather than efficiency. Highly efficient systems are usually very rigid and cannot effectively respond to changing incentives and inhibitions. Personal commitments from independent agents underlie relationships rather than contractual obligations due to the agents' independence. Heterarchical networks among agents provide the dominant social mechanism despite the typical existence of a nominal hierarchy.

Hierarchical organizations are usually designed in terms of roles, reporting structures, etc. Heterarchical systems are much more likely to self-organize. This is particularly true for organizations of professionals who also have strong affiliations with professional associations. A good example is physicians, whose professional affiliations have strongly driven the evolution of the healthcare system (Rouse & Cortese, 2010).

The characteristics summarized in Table 1.1 have significant impacts on how such organizational systems address fundamental change or transformation. The leaders cannot simply command change. Instead, change has to emerge organically in terms of agents' responses to new incentives and inhibitions. Transformation will emerge from agents' collective self-organization.

A useful construct in this regard is the notion of "tipping points," popularized by Gladwell (2000). Within the context of organizational transformation, the idea is that a small change can propagate and gain momentum across the heterarchical organizational network and lead to bottom-up discussions and desires for fundamental change. In other words, top-down "push" for change is replaced by bottom-up "pull" for change.

In healthcare, for example, there is a pending change from fee for service to payment for health outcomes (Rouse & Serban, 2014). While this is far from fully implemented, it has caused healthcare provider organizations to study carefully how they will need to adapt to such a fundamental change. More specifically, it has caused serious study of what does and does not affect health outcomes.

One approach to understanding possible tipping points involves mathematical and computational modeling of the enterprise (Rouse, 2015). These models can be useful in the exploration of leading indicators of the different tipping points and in assessing potential ways to accelerate desirable outcomes as well as mitigations for undesirable outcomes. Such models can be embodied in interactive visualizations called "policy flight simulators" that enable stakeholders to computationally experiment with alternative futures (Rouse, 2014).

UNIVERSITIES AS COMPLEX ADAPTIVE SYSTEMS

The construct of complex adaptive systems can help to understand and enable fundamental change both within and across universities. As a first step in this direction, it is useful to discuss how the organizational phenomena discussed earlier are manifested in university environments.

Nonlinear, Dynamic Behavior

Linear behavior means a system responds proportionately to stimuli. Nonlinear behavior means disproportionate responses. University communities typically under-respond in the sense that, for the most part, they ignore stimuli. Faculty members tend to be oblivious to most things external to their specialty. Occasionally, they overrespond, for instance, to offhand comments by the administration that can be interpreted as being less than fully committed to some faculty initiative. A typical result is a barrage of emails from all corners of the community.

Independent Agents

Faculty members embrace academic freedom in a broad sense. This includes freedom in what they research and how they research it. Some faculty members simply go in their office, close the door, and immerse themselves in their scholarly pursuits. More broadly, faculty members feel free to openly oppose organizational endeavors. I have experienced many instances of faculty members investing significant energy to undermine initiatives of senior administrators. Occasionally cabals of faculty members emerge to support or oppose a cause.

Goals and Behaviors That Differ or Conflict

Faculty disagreements are common and occasionally heated. Disrespect across disciplines is a good example. A portion of the faculty will assert that the only fundamental research is that which involves axioms and theorems. A different portion of the faculty will assert that theory and experimentation are the essence of science. If these two groups are in the same academic department, there can be much conflict, for example, about requirements for the Ph.D. degree. I have occasionally defused such conflicts, at least temporarily, by noting that we not only need Newtons and Einsteins; we also need Darwins.

Intelligent and Learning Agents

Most faculty members were good students. They were bright, did their homework, and got top grades. Thus, they are intellectually intelligent, although not always emotionally intelligent. They learn how their university system works and they learn how to work the system to achieve their desired ends. Of course, as indicated earlier, there is seldom universal agreement on the ends. The result is a lot of smart, usually well-intended, people working at odds with each other. University administrators call this situation "herding cats." It often results in creative but hopefully not destructive chaos and substantially impedes progress.

Self-Organization

There are two types of organization of interest. As shown in Figure 1.1, there are colleges, schools, and departments that invest in processes to support practices. There are also self-organized interest groups within and sometimes across departments. In addition, as depicted in Figure 1.2, there are discipline-specific organizations across universities that, as discussed earlier, define research agendas and vet sources of funding and publication outlets. These groups self-organize independent of the hierarchies of the universities shown in Figure 1.1.

No Single Point(s) of Control

The president and board of trustees (or board of regents) are nominally in control of a university. They approve hiring, promoting, and tenuring of faculty members. They are the authority that grants degrees. They decide about operating and capital budgets. However, they have little control over the daily activities of the many independent agents in the university community. In this sense, the president has much more influence than power. If, for instance, a subset of the faculty disagrees with the president's vision for the university, they can drag their heels and wait until he or she leaves office.

Implications

Referring to Table 1.1, universities are better led than managed. The leadership needs to articulate and gain support for a vision of the university's future. They also need to create and sustain a set of incentives and inhibitions that motivate the independent agents to personally commit to pursuit of the vision. There need to be agreed-upon outcomes that successful pursuit of the vision will yield.

The organization needs to be sufficiently agile to pursue the vision and create the desired outcomes. Both the hierarchy and heterarchy need to be agile in the sense of having the orientation and resources needed to make the changes necessary to successfully pursue the vision. To some extent, the requisite agility can be designed into the organization. However, inclinations to self-organize can yield creative and highly effective modifications to any initial design.

OVERVIEW OF CHAPTERS

Chapter 1: Introduction and Overview

This chapter begins by providing a characterization of the overall architecture of a university enterprise and its relationship to other university enterprises. These diagrams provide a framework for discussing the overall approach of this book as contrasted with the enormous body of literature on higher education. The nature of complex enterprises is then considered, with emphasis on complex adaptive systems. The specific attributes of universities that reflect the characteristics of complex adaptive systems are elaborated. Finally, an overview of the book is provided.

Chapter 2: Evolution of the Research University

This chapter begins in Europe where universities as independent institutions first emerged almost 1000 years ago, followed by consideration of early universities in America. In the early nineteenth century, Wilhelm von Humboldt transformed German higher education by integrating teaching and research. This leads to discussion of the Morrill Land-Grant Acts in the United States that provided resources that enabled adoption of a version of the Humboldt model, with less hierarchy and more participation.

The central role of World War II in advancing science and engineering is next discussed. The war also prompted the recognition that engineering education needed stronger roots in science and mathematics. Pursuit of engineering science led to great advances but also sometimes led to overemphasis on mathematics and intense specialization. These tendencies have served as both strengths and weaknesses. The many challenges faced by today's research universities are discussed—they stem in part from these forces.

Chapter 3: Mission and Structure

This chapter begins by reviewing several university mission statements. They typically have much in common with regard to education, research, and service. A substantial problem in recent decades has been mission creep, whereby universities attempt to provide value in varying ways to different constituencies. This inevitably leads to greater costs, which are mostly recouped via tuition increases despite the fact that many of the new activities do not directly benefit those paying tuition.

This chapter next addresses the multilevel structure of universities. Education, research, and service practices occur in the context of processes that provide capabilities that enable practices. These capabilities are created via investments by departments, schools, colleges, and campuses. These investments are motivated by the "rules of the game" created and sustained by the overall ecosystem.

Finally, this chapter discusses how the traditional structure of universities does not align well with the mission and needs of interdisciplinary research centers. There are inherent conflicts between traditional discipline-oriented departments, schools, and colleges and the crosscutting nature of interdisciplinary research centers. This poses challenges for faculty members who are attracted to larger problems but also want to be successful in securing tenure and promotions.

Chapter 4: Leadership and Governance

This chapter considers the role of leadership as enterprises seek to change, including the importance of how leaders spend their time. The negative impacts of stewards of the status quo are elaborated. Experiences of leading research centers are considered. Examples of good and bad leadership are discussed.

Governance is considered in terms of governing boards, administration and faculty, and other players. Numerous vignettes of governance are discussed. Decision making in universities clearly exhibits the characteristics of complex adaptive systems. Many independent agents influence the process and much time is typically consumed getting to a decision.

Chapter 5: Administration

This chapter addresses the nature of administration in terms of managing and sustaining a university. The corporatization of academia has resulted in enormous growth of the number of administrators and administrative staff and costs of these personnel. A relatively simple model is proposed for projecting these costs.

This chapter also considers several core administrative functions, including performance evaluation, conflict management, compliance and abuse, and marketing and communications. The costs of conforming to various administrative practices are discussed.

Chapter 6: Money and Space

This chapter addresses the economics of higher education, including the value of education, the impacts of government subsidies, and the possibilities of a higher education bubble. The costs of higher education are discussed in terms of cost analyses, indirect costs, staffing patterns, and student and institutional debt. Consideration of revenue sources includes discussion of tuition, government dependencies, and philanthropic fundraising. An overall economic model of a university is presented. Finally, the challenging issue of space is addressed.

Chapter 7: Promotion and Tenure

This chapter addresses the nature and process of promoting and tenuring university faculty members. The nature and roles of faculty are discussed in terms of academic disciplines, faculty impact, tenure-track versus nontenure-track positions, availability of positions, and faculty turnover. The nature of tenure decisions is explored. I relate my experiences serving on several promotion and tenure committees. Finally, a model of tenure decision making is introduced that is integrated into an overall model in Chapter 12.

Chapter 8: Education Programs

This chapter addresses the current and future student populations that higher education will serve. Both science, technology, engineering, and mathematics (STEM) challenges and evolving student inclinations and preferences are discussed. Challenges to higher education's value proposition are reviewed. Types of degree programs and the design and delivery of curricula and courses are considered, including alternative approaches to delivery. I relate a variety of my teaching experiences. Finally, a workforce model is introduced and a range of staffing trade-offs is discussed.

Chapter 9: Research and Intellectual Property

This chapter addresses the challenges of research in terms of the role of peer review of publications and proposals, bibliometrics associated with citations, and difficulties securing funding. A variety of research experiences are related. A model of research is presented that predicts probabilities of articles being accepted, expected citations after publication, and probabilities of research proposals being funded. This model is a central element of the integrated model discussed in Chapter 12.

The last major section of this chapter addresses intellectual property, which is often one of the outcomes of research. In some cases, this takes the form of patents, but more often it is embodied in the know-how of the researchers. Either form of intellectual property may result in spin-off businesses that create jobs, revenue, profit, and perhaps public stock offerings. I relate several of my experiences in launching and growing spin-offs.

Chapter 10: Rankings and Brand Value

This chapter addresses several ranking schemes that emphasize different attributes of a university. Georgia Tech is used as an example to illustrate the types of initiatives needed to improve rankings. Statistical analyses of rankings of one program—industrial/manufacturing/systems engineering—are reported. The best predictors of a program's ranking are last year's ranking plus the number of faculty.

The brand value of a university is discussed as a proxy for rankings. An index of brand value is presented that is a weighted sum of number of articles published, number of citations received, and h-index, totaled across all faculty members of an institution. An example is used to illustrate the nuances of the brand value index as affected by research sponsors and publication outlets chosen by faculty members.

Chapter 11: Transformation Scenarios

In this chapter, forces for change are discussed in terms of costs and benefits, globalization, and technology. Organizational change in higher education is then addressed, including concepts and principles drawn from domains other than higher education. Four alternative scenarios for the future of higher education are next elaborated. These scenarios provide the basis for considering transformation of academia. The chapter concludes with a discussion of historical perspectives on how change happens. This sets the stage for exploring the future.

Chapter 12: Exploring the Future

This chapter brings together all the pieces of the puzzle elaborated in this book. The elements of the education and research enterprise are integrated to enable projecting the likely consequences of several rather disparate scenarios of the future of the academic enterprise. We cannot predict what mix of these scenarios will actually emerge. However, we can argue that universities need strategies and investments that enable robust responses to whatever mix emerges.

Fundamental change is in the offing. Higher education cannot remain the poster child for runaway costs. We need a healthy, educated, and productive population that is competitive in the global marketplace. If the population is not educated, it will not be healthy. If the population is not productive, it will not be competitive. The pieces all fit together. We have to make it happen.

REFERENCES

Christenson, C.M., & Eyring, H.J. (2011). *The Innovative University: Changing the DNA of Higher Education from Inside Out.* San Francisco, CA: Jossey-Bass.

DeMillo, R.A. (2011). *Abelard to Apple: The Fate of American Colleges and Universities.* Cambridge, MA: MIT Press.

Gladwell, M. (2000). *The Tipping Point: How Little Things Can Make a Big Difference.* Boston, MA: Little Brown.

Lombardi, J.V. (2013). *How Universities Work.* Baltimore, MD: Johns Hopkins University Press.

Poli, R. (2013). A note on the difference between complicated and complex social systems. *Cadmus, 2* (1), 142–147.

Rouse, W.B. (2000). Managing complexity: Disease control as a complex adaptive system. *Information Knowledge Systems Management, 2* (2), 143–165.

Rouse, W.B. (2008). Healthcare as a complex adaptive system: Implications for design and management. *The Bridge, 38* (1), 17–25.

Rouse, W.B. (2014). Human interaction with policy flight simulators. *Journal of Applied Ergonomics, 45* (1), 72–77.

Rouse, W.B. (2015). *Modeling and Visualization of Complex Systems and Enterprises: Explorations of Physical, Human, Economic, and Social Phenomena.* Hoboken, NJ: Wiley.

Rouse, W.B., & Cortese, D.A. (Eds.). (2010). *Engineering the System of Healthcare Delivery.* Amsterdam, The Netherlands: IOS Press.

Rouse, W.B., & Serban, N. (2014). *Understanding and Managing the Complexity of Healthcare.* Cambridge, MA: MIT Press.

Snowden, D.J., & Boone, M.E. (2007). A leader's framework for decision making. *Harvard Business Review,* November, 69–76.

2

EVOLUTION OF THE RESEARCH UNIVERSITY

This book is about how academia works, why it works these ways, and where the university enterprise is headed. To address these questions in meaningful ways, we must start with a firm understanding of the evolution of research universities. This chapter is intended to provide this understanding.

We begin where universities as independent institutions first emerged—in Europe almost 1000 years ago. These institutions focused on teaching, not research. However, their legacy of organizational structure and governance models has strongly impacted the ways contemporary universities organize and operate.

We next consider early universities in America. For a bit less than the first half of the four centuries of universities in America, teaching was also the focus. People were classically educated to serve in the clergy and government. Science and engineering were not yet seen as central to universities.

This all changed when Wilhelm von Humboldt transformed German education. He integrated teaching and research while insisting that both be based on unbiased knowledge and analysis. The Morrill Acts in the United States provided resources that prompted adoption of a version of the Humboldt model, with less hierarchy and more participation. For the last 150 years, research universities in the United States built on this model have prospered.

World War II played a central role in advancing science and engineering. Vannevar Bush was visionary in how he saw that academia and government should interact. His vision was instrumental in the formation of the National Science

---

Universities as Complex Enterprises: How Academia Works, Why It Works These Ways, and Where the University Enterprise Is Headed, First Edition. William B. Rouse.
© 2016 John Wiley & Sons, Inc. Published 2016 by John Wiley & Sons, Inc.

Foundation (NSF). The war also prompted the recognition that engineering education needed stronger roots in science and mathematics.

Pursuit of engineering science not only led to great advances but also sometimes led to overemphasis on mathematics and intense specialization. These tendencies have served as both strengths and weaknesses. Many of the challenges faced by today's research universities stem from these forces. Fortunately, the historical summary provided by this chapter will help to meaningfully formulate productive and valuable ways forward.

EARLY UNIVERSITIES IN EUROPE

The ancient Greeks and Romans educated their citizens, as least the upper classes. Families were responsible for how education was delivered. However, there were no educational institutions per se. Such institutions did not emerge until the beginning of the last millennium.

In this section, I review five of the earliest institutions of higher education, in England, France, and Italy. We will see, for example, that important elements of contemporary universities' organizational structures can be traced back to institutions formed almost 1000 years ago. We will also see how important governance issues were addressed.

University of Bologna

In 1088, teaching began in Bologna, independent of religious schools (Bologna, 2015).[1] Teachers of grammar, rhetoric, and logic applied themselves to law. Debates about law were central at that time to resolution of the relative roles of church and state. By the fourteenth century, there were scholars of other domains of study, including arithmetic, astronomy, logic, and medicine.

The Roman Emperor Frederick Barbarossa protected scholars from intrusion by political authorities. The university was legally declared to be independent of other powers, a fundamental facet of the emerging European university. After the death of the emperor, the university often fought to retain autonomy, while political powers continually tried to influence it.

By the thirteenth century there were over 2000 students in Bologna. The students organized themselves by their country of region of origin. These organizational entities were termed "nations," and the collaborative of nations was called universitas. The students initially paid the teachers gifts rather salaries. These gifts were eventually transformed into actual salaries. However, students did not always pay, and the city had to contribute.

---

[1] The references noted in these subsections refer to the histories of these universities as articulated on their websites, which are included in the reference list.

University of Paris

The University of Paris emerged in 1150 as a grouping of several schools that were located near each other (Paris, 2015). These schools provided instruction at three levels: baccalauréat (grammar, dialectics, rhetoric), license (arithmetic, geometry, astronomy, music), and doctorate (medicine, canon law, theology). Their success soon made it necessary to create an organized structure. King Philippe Auguste agreed to provide the teachers and students with suitable living conditions and to guarantee—with diplomas—the quality of their education, which was becoming a means of social mobility.

The schools' organization system was defined in accordance with two major principles. Teachers and students were grouped together in a community called universitas that was governed by statutes setting out the rules of communal life within a shared system of education. The second principle was that of self-government, which was guaranteed by the king, confirmed by papal legate, and later endorsed by papal bull from Pope Gregory IX. Thus, the university became a legal entity, with top-down rules and governance, unlike the bottom-up student-driven organization of Bologna.

The students at Paris also formed four "nations" depending on their country or region of origin. Schools emerged from the formation of "faculties" in liberal arts, medicine, canon law, and theology. By the end of the fifteenth century, the University of Paris had become the biggest cultural and scientific center in Europe, attracting some 20,000 students.

University of Oxford

In 1167, King Henry II banned English students from attending the University of Paris (Oxford, 2015). This led to rapid development of teaching at Oxford. By 1201, the university was headed by a chancellor, and in 1231 Oxford was recognized as a universitas or corporation. The oldest of Oxford's colleges—University, Balliol, and Merton—were established between 1249 and 1264. Less than a century later, the university had achieved eminence and won praise for its antiquity, curriculum, doctrine, and privileges. Oxford is ranked as one of the top research universities in the world.

Oxford University Press (OUP, 2015) is the largest university press in the world and the second oldest after Cambridge University Press. The first book was printed in Oxford in 1478, less than four decades after Gutenberg introduced the printing press in 1440. The university was involved with several printers in Oxford over the next century, although there was no formal university press. In 1586 the University of Oxford's right to print books was recognized in a decree from the Star Chamber—an English court of law at the Royal Palace of Westminster. A charter subsequently secured by Archbishop Laud from King Charles I entitled the university to print all manner of books.

University of Padua

The University of Padua was established in 1222, after a group of students and teachers moved there from Bologna (Padua, 2015). They organized an independent body of scholars, who were grouped according to their place of origin into nations.

Students approved statutes, elected the rector or chancellor, and chose their teachers, who were paid with money the students collected. Freedom of thought in study and teaching became a central facet of Padua.

The university's introduction of empirical and experimental methods, together with the teaching of theory, marked the dawn of a golden age. In the sixteenth and seventeenth centuries, Padua became a workshop of ideas and the home to figures that changed the cultural and scientific history of humanity. Padua also created the world's first university botanical garden and a permanent anatomical theater. In 1678, Elena Lucrezia Cornaro Piscopia became the first woman in the world to be awarded a university degree.

University of Cambridge

In 1209, scholars taking refuge from hostile townsmen in Oxford migrated to Cambridge, a town bridging the River Cam since 875 (Cambridge, 2015). They initially lived in lodgings in the town, but eventually housing was acquired with a master in charge of the students. By 1226 the activity was sufficient to require an organization, represented by a chancellor. Regular courses of study were designed and taught by scholars belonging to the organization.

King Henry III took the scholars under his protection as early as 1231 and arranged for them to be sheltered from exploitation by city landlords. He provided them a monopoly on teaching. Only those enrolled under the tuition of a recognized master were allowed to remain. Similar to the other universities described here, the students at Cambridge first studied the "arts"—grammar, logic, and rhetoric—followed later by arithmetic, astronomy, geometry, and music, leading to the degrees of bachelor and master.

There were no professors. Masters who had passed the course and who had been approved or licensed by the universitas conducted the teaching. The teaching took the form of reading and explaining texts. The examinations were oral disputations. Over time, of course, Cambridge became known for many famous faculty members and is ranked as one of the world's top research universities.

Summary

The five institutions reviewed in the previous text were the first to move beyond religious teaching institutions. They were formed to teach law, medicine, and theology. Mathematics and science were added later. Engineering did not emerge as a discipline warranting formal education until the mid to late eighteenth century.

Much of contemporary university organizational structures can be traced to these five institutions. The constructs of departments, schools, and colleges were defined then, as was the typical hierarchical structure among these organizational entities. These institutions also defined the roles of chairs, deans, and chancellors or rectors.

University governance varied, ranging from the bottom-up student-led Bologna to the top-down government-led Paris. Similarly, students provided the resources in Bologna, while government provided resources in Paris. Padua was similar to Bologna, while Oxford and Cambridge were more like Paris with regard to these issues.

Bottom-up governance, whether by faculty or students, affected curricula and requirements for credentials. Top-down governance, whether by the administration or government, set or at least agreed to the overall policies and procedures of the institution and addressed facilities and other resources. Other political interests occasionally tried to intervene but were often unsuccessful.

All in all, the legacies of these institutions have been immense. Organizational structures and governance processes have been passed down through the centuries. Contemporary faculty members and administrators tend to view these principles as inviolate. I revisit these principles—as well as their consequences—in several later discussions.

EARLY UNIVERSITIES IN AMERICA

The earliest universities in America included Harvard (1636), Yale (1701), Pennsylvania (1740), Princeton (1746), Columbia (1754), and Brown (1764). The other member of the Ivy League—Cornell—was founded a century after Brown and is later discussed. With the exception of Pennsylvania, all of these institutions were founded by religious denominations.

Religious communities founded early colleges for the purpose of promoting and maintaining their particular religious perspective. Their educational mission focused on educating clergy and public officials. Those who aspired to careers other than clergy or government positions sought practical training emphasizing reading, writing, and mathematics as necessary tools of business, including training in the practical application of these topics to trades such as shipbuilding, trading, or farming.

After the Revolutionary War, a college degree became a status symbol for wealthy citizens who would send their sons to be educated at Harvard, Yale, or Princeton with no pretense that this education was preparing their sons for lives of service to church or government. Schools of law, medicine, engineering, and finance and accounting emerged in the early nineteenth century. Such professional schools were alternatives to college and did not provide a classical Latin education.

While the aforementioned Ivy League universities eventually became leading research universities, they were not founded for this purpose, and it took one or two centuries for them to also embrace this mission. The stepping-stone to this broader mission was a German educational innovation.

HUMBOLDT'S INNOVATION

Wilhelm von Humboldt and a group of colleagues founded the University of Berlin in 1810 (Berlin, 2015). Humboldt was a Prussian philosopher, government official, and diplomat. As an official in the Interior Ministry, he reformed the Prussian school and university system to create an educational system that integrated the natural sciences, social sciences, and humanities. Research and teaching were combined and based on unbiased knowledge and analysis. Students were allowed to choose their own course of study.

"Humboldt's model was based on two ideas of the Enlightenment: the individual and the world citizen. Humboldt believed that the university (and education in general, as in the Prussian education system) should enable students to become autonomous individuals and world citizens by developing their own reasoning powers in an environment of academic freedom. Humboldt envisaged education in a broad sense, which aimed not merely to provide professional skills through schooling along a fixed path but rather to allow students to build individual character by choosing their own way" (Anderson, 2004; Berlin, 2015).

These principles of Wilhelm von Humboldt and his contemporaries soon became general practice throughout the world. A new era of university and academic research had begun. Indeed, contemporary research universities date back only to Humboldt's University of Berlin in the early nineteenth century. As noted earlier, before that universities were devoted to preparing professionals for law, medicine, and theology (Altbach, 2011).

Japan and the United States were among the most enthusiastic adopters of the German model. The US model differed from the German in emphasizing service to society, being organized more democratically, having participatory governance, and being managed by deans and presidents appointed by trustees. By the middle of the twentieth century, the United States was the gold standard for research universities (Altbach, 2011).

In 1949, the University of Berlin was renamed Humboldt University after Wilhelm von Humboldt and his brother, naturalist Alexander von Humboldt.

MORRILL ACTS

The Morrill Land Grant Act of 1862 profoundly affected research universities in the United States. Before discussing this impact, it is important to summarize innovations in engineering education prior to the Morrill Act.

The US Military Academy (USMA) at West Point was established by President Jefferson in 1802 and began operation in 1803. "Colonel Sylvanus Thayer served as Superintendent from 1817 to 1833. He upgraded academic standards, instilled military discipline and emphasized honorable conduct. He made civil engineering the foundation of the curriculum. For the first half of the nineteenth century, USMA graduates were largely responsible for the construction of the bulk of the nation's initial railway lines, bridges, harbors and roads" (West Point, 2015).

The Rensselaer School was established in Troy, New York, in 1824 by Stephen van Rensselaer, along with educator and scientist Amos Eaton, "for the purpose of instructing persons…in the application of science to the common purposes of life." Rensselaer Polytechnic Institute (RPI) is "…the first school of science and school of civil engineering, which has had a continuous existence, to be established in any English-speaking country" according to Palmer C. Ricketts in his preface to the second edition of his *History of Rensselaer Polytechnic Institute* (1914). In 1833 the school became the Rensselaer Institute, and in the 1850s its purpose was broadened to become a polytechnic institution. The institute's name was changed in 1861 to Rensselaer Polytechnic Institute (RPI, 2015).

Thomas Jefferson founded the University of Virginia in 1819. He envisioned a new kind of university, one dedicated to educating leaders in practical affairs and public service rather than for professions in the classroom and pulpit exclusively. It was the first nonsectarian university in the United States and the first to use the elective course system. The School of Engineering and Applied Science began as the School of Civil Engineering in 1836. It was the first engineering school in the south. Charles Bonnycastle and William Barton Rogers, who in 1864 would become founding president at the Massachusetts Institute of Technology (MIT), established the school (Virginia, 2015).

The Morrill Land Grant Act of 1862 led to transformation, or Americanization, of the Humboldt model for research universities. As indicated earlier, the American version of the German model was organized more democratically, had participatory governance, and was managed by deans and presidents appointed by trustees. In contrast, the German model invested ultimate power in the senior faculty members who elected the chancellor or rector from among their peers.

The Morrill Land-Grant Acts, named for Representative Justin Smith Morrill of Vermont, are US statutes that enabled the creation of land-grant colleges that would teach agriculture and engineering. The purpose of the land-grant colleges was "Without excluding other scientific and classical studies and including military tactics, to teach such branches of learning as are related to agriculture and the mechanic arts, in such manner as the legislatures of the States may respectively prescribe, in order to promote the liberal and practical education of the industrial classes in the several pursuits and professions in life." These acts include the Morrill Act of 1862 and the Morrill Act of 1890.

The Morrill Act of 1862 allocated 30,000 acres of federal land based on the number of senators and representatives each state had in Congress. This land, or the proceeds from its sale, was to be used toward establishing and funding the requisite educational institutions. If the federal land within a state was insufficient to meet that state's land grant, the state was issued "scrip," which authorized the state to select federal lands in other states. For example, New York selected valuable timberland in Wisconsin to fund Cornell University. The resulting management of this scrip by Cornell yielded one-third of the total grant revenues generated by all the states, even though New York received only one-tenth of the 1862 land grant. Overall, the 1862 Morrill Act allocated 17,400,000 acres of land.

The act was first proposed in 1857 and was passed by Congress in 1859, but President James Buchanan vetoed it. Morrill resubmitted the act in 1861 with the amendment that the proposed institutions would teach military tactics as well as agriculture and engineering. Aided by the secession of many states that did not support the act, this reconfigured Morrill Act was signed into law by President Abraham Lincoln on July 2, 1862.

A second Morrill Act in 1890 was aimed at the former Confederate states. This act required each state to show that race was not an admissions criterion, or else to designate a separate land-grant institution for persons of color. Among the seventy colleges and universities that benefited from the Morrill Acts are several historically black colleges and universities (HBCUs). Though the 1890 Act granted cash instead of land, it established that colleges under the 1890 Act had the same legal standing as the 1862 Act colleges. Hence, the phrase land-grant college properly applies to both groups.

The land grants, especially the proceeds from the land, enabled enormous investments in what became flagship state universities. The University of Arizona, University of Florida, University of Illinois, University of Maryland, Ohio State University, Pennsylvania State University, University of Wisconsin, and, all told, more than 70 institutions became land-grant colleges—most of them are now universities. This, of course, provided a huge boost to education in agriculture and engineering.

MIT (1865) and Cornell University (1865) were two early land-grant universities. Many other institutes of technology were soon founded including Worcester Polytechnic Institute (1865), Stevens Institute of Technology (1870), Georgia Institute of Technology (1885), California Institute of Technology (1891), and Carnegie Institute of Technology (1900), which later became Carnegie Mellon University. As noted in the previous text, Rensselaer Institute was renamed in 1861 to Rensselaer Polytechnic Institute.

Founding of these types of institutions contributed to what I have termed the "Century of Innovation"—a period from roughly 1840 to 1975, so 135 years, that saw a transition from wooden sailing ships to electric railways, computers, space travel, and the Internet (Rouse, 2014). The land-grant colleges and these institutes of technology provided the talent and contributed to many of the inventions that enabled these wonderful innovations.

BUSH AND NSF

The pre-World War II twentieth century provided a hotbed of research in physical sciences and mathematics. Physics and computing are of particular note. However, the modern research university, particularly in the United States, emerged following World War II. Vannevar Bush, a twentieth-century leader in engineering and science, was instrumental in defining the vision.

Bush articulated the central principles in *Science: The Endless Frontier* (Bush, 1945):

- The federal government shoulders the principal responsibility for the financial support of basic scientific research
- Universities—rather than government laboratories, nonteaching research institutes, or private industry—are the primary institutions in which this government-funded research is undertaken
- Although the federal budgetary process determines the total amount available to support research in various fields of science, most funds are allocated not according to commercial or political considerations but through an intensely competitive process of review conducted by independent scientific experts who judge the quality of proposals according to their scientific merits alone

Bush's leadership led to the establishment of the NSF via the National Science Foundation Act of 1950. NSF's stated mission is "To promote the progress of science; to advance the national health, prosperity, and welfare; and to secure the national defense." The NSF is certainly not the primary government agency for the funding of basic science and engineering, as many envisioned in the aftermath of World War II. Specialized agencies such as the National Institutes of Health (medical research), the US Atomic Energy Commission (nuclear and particle physics), National Aeronautics and Space Administration (space science), and the Defense Advanced Research Projects Agency (defense-related research) dominate support for these major research areas. Nevertheless, the NSF funds approximately 20% of all federally supported *basic research* conducted by the US colleges and universities. Thus, Bush's vision has, at least in part, long been realized.

Perhaps not surprisingly, Bush's home university, the MIT, was very successful in adopting his principles. James Killian, MIT president from 1949 to 1959, notes that "From MIT's founding, the central mission had been to work with things and ideas that were immediately useful and in the public interest. This commitment was reinforced by the fact that many faculty members had had during the war direct and personal experience in public services" (Killian, 1985, p. 399).

He reports that MIT's relationship with the federal government reached new heights with World War II:

- MIT took on critical challenges, for example, the Sage missile defense system and the Whirlwind computing project
- Faculty and alumni serving in important advisory roles in the federal government
- Faculty, including two MIT presidents, served in senior executive positions, on leave from MIT

As a consequence, MIT became and remains a national resource, perhaps the key player in "big science." In the process, MIT was transformed into a university. This was facilitated by several factors (Killian, 1985):

- A single, unfragmented faculty in consort with one central administration
- Close articulation of research and teaching of basic science and applied science
- Continuous spectrum of undergraduate and graduate studies
- Mobility of ideas resulting from the high permeability of the boundaries of both departments and centers
- The extensive interconnection of its buildings

MIT and a handful of other leading institutions such as the University of California at Berkeley, California Institute of Technology, University of Illinois, and Stanford University led the way defining the nature and "rules of the games" for research universities.

ENGINEERING SCIENCE

Scientists rather than engineers achieved many of the key engineering accomplishments during World War II. Conventional engineering education at that time had not prepared engineers to deal with phenomena such those as associated with radar. These types of experience led to demands for more science and mathematics in engineering education (NRC, 1985).

The push to include more science in engineering was formalized by the publication of the *Report on Evaluation of Engineering Education* (ASEE, 1955), which recommended strengthening of the sciences in graduate engineering program as well as developing superior faculty members.

The President's Advisory Committee on Science report on *Meeting Manpower Needs in Science and Technology* (PSAC, 1962) cast similar recommendations in the context of national security and competitiveness, prompted in part by the Russians beating the United States into space by 4 years with Sputnik.

A third report, *Goals for Engineering Education* (ASEE, 1968), reinforced these recommendations, although the report's projections conflicted with the reality of the 1969–1971 retrenchment of large aerospace enterprises due to the greatly reduced spending for military acquisitions. By the 1980s, however, the projected oversupply of engineers had disappeared.

The legacy of the emphasis on engineering science has been enormous. It has led to great emphasis on mathematics and intense specialization. Many faculty members will avoid addressing problems that cannot be mathematically formulated. The result has been much research on "toy" problems that can be tightly formulated and amenable to solutions that can be proven to be optimal. In the extreme, the goal becomes one of proving theorems rather than solving realistic problems.

Intense specialization has created numerous subdisciplines and a proliferation of journals focused solely on these subdisciplines. Seams among subdisciplines make it very difficult to address larger problems. Seams between disciplines, for example, physical and social sciences make it very difficult to formulate transdisciplinary problems and solutions.

Many commentators have articulated these limitations of academia and argued for the importance of interdisciplinary collaborative teams for addressing problems such as healthcare delivery or urban resilience. Faculty members know, however, that such endeavors are fraught with risks such as publications coming too slowly and credit being spread too thinly. This issue is discussed in more depth later in this book.

TODAY'S RESEARCH UNIVERSITIES

Science and technology has become central to our economy. As Richard Levin, former president of Yale University, indicates, "Competitive advantage based on the innovative application of new scientific knowledge—this has been the key to American economic success for at least the past quarter century" (p. 88). He asserts that the success of this system is evident: The United States accounts for 33% of all scientific publications and has won 60% of Nobel Prizes, and its universities account for 73% of papers cited in US patents (Levin, 2003).

However, it is not clear that this traditional model is sustainable. James Duderstadt (2000), former president of the University of Michigan, summarizes several areas of concern identified in an NSF study:

- Public support has eroded with continual decline throughout the 1990s
- Limits on indirect costs have resulted in cost shifting
- The focus on research funding has changed the role of the faculty
- Increased specialization has changed the intellectual makeup of academia

He argues that the real issue is a shifting paradigm for universities. National priorities have changed, although recent security concerns have moderated this trend. The disciplines have been deified, yielding a dominance of reductionism. This presents a challenge for interdisciplinary scholarship, particularly in terms of valuing a diversity of approaches and more flexible visions of faculty career paths. At the same time, undergraduate education is receiving increased attention, as have cultural considerations that, he cautions, tend to encourage "belongers" rather than "doers."

Ruminating on the roles of publicly supported research universities, Duderstadt suggests several possibilities for strategies that universities can pursue as a response to current challenges:

- Isolation: Stick with prestige and prosperity, for example, MIT, Caltech, Princeton, and Chicago
- Pathfinders: Participate in experiments creating possible futures for higher education
- Alliances: Allying with other types of educational institutions
- Core-in-Cloud Models: Elite education and basic research departments surrounded by broader array of entities

Derek Bok, relatively recent president of Harvard University, addresses the future of universities in light of many recent trends (Bok, 2003). He is particularly concerned with the commercialization of the university in response to a plethora of "business opportunities" for universities. He notes that "Increasingly, success in university administration came to mean being more resourceful than one's competitors in finding funds to achieve new goals. Enterprising leaders seeking to improve their institution felt impelled to take full advantage of any legitimate opportunities that the commercial world had to offer" (p. 15). He argues that this increased focus on commercialization may jeopardize the focus on education and learning.

Bok recognizes that this shift is nevertheless taking place. He cautions, however, that universities typically face several challenges that can hinder entrepreneurial aspirations. Bok summarizes these challenges: "On three important counts, the environment in most research universities does not do enough to encourage the behaviors needed for the sake of the students, the society, and the well-being of the institution itself" (pp. 23–24).

- Efficiency: "University administrators do not have as strong incentives as most business executives to lower costs and achieve greater efficiency" (p. 24)
- Improvement: "A second important lesson universities can learn from business is the value of striving continuously to improve the quality of what they do" (p. 25)
- Incentives: "Left to itself, the contemporary research university does not contain sufficient incentives to elicit all the behaviors that society has a right to expect" (p. 28)

These seem like reasonable challenges, at least for businesses. However, Bok argues that "Leading a university is also a much more uncertain and ambiguous enterprise than managing a company because the market for higher education lacks tangible measurable goals by which to measure success" (p. 30). Further, he asserts that "Presidents and deans are ultimately responsible for upholding basic academic values but they are exposed to strong conflicting pressures that make it hard for them to carry out this duty effectively" (p. 185).

We would expect that market forces would resolve these pressures. However, Bok reasons that "Neither the profit motive nor the traditional methods of the research university guarantee that faculties will make a serious, sustained effort to improve their methods of instruction and enhance the quality of learning on their campuses" (p. 179). In other words, we cannot expect an organically based transformation of academia, despite financial and social forces for fundamental changes. There is a fundamental tension between what is naturally happening in research institutions (i.e., increased focus on the external viability of research) and the way in which this is being managed, or not managed, within the same universities. The lack of attention and process in the midst of this evolution could halt the progress and risk the outcomes of the changes.

The challenges outlined by these former leaders of some of our best research universities provide motivation for much of the discourse in this book. As noted in Chapter 1, complex adaptive systems are difficult to change, and universities, as exemplars of such systems, can be enormously creative in finding ways to perpetuate the status quo. The concepts, principles, models, and methods discussed in the following chapters will help to identify tipping points where productive and valuable change is possible.

CONCLUSIONS

This chapter has provided a very quick tour of 1000 years of academia. We began in Europe where issues of organizational structure and governance models were addressed in the context of teaching institutions. We next considered early universities in America where teaching was also the primary objective.

This all changed when Wilhelm von Humboldt transformed German education. He integrated teaching and research while insisting that both be based on unbiased knowledge and analysis. The Morrill Acts in the United States provided resources that prompted adoption of the Humboldt model, with a bit less hierarchy and more participative governance.

World War II played a central role in advancing science and engineering. Vannevar Bush's vision was instrumental in the formation of the NSF. The war also prompted the recognition that engineering education needed stronger roots in science and mathematics. Pursuit of engineering science led to great advances, but this emphasis resulted in both strengths and weaknesses. Many of the challenges faced by today's research universities stem from these forces.

This 1000-year tour enables understanding of the evolution of research universities. We know why things are the way they are, at least at a high level. To entertain change, however, we need to address the organization and operation of universities in much greater depth. The rest of this book pursues this depth focused on how best to understand and address avenues of change and impediments to change in the complex academic enterprise.

REFERENCES

Altbach, P.G. (2011). The past, present, and future of the research university. In P.G. Altbach & J. Salmi, Eds., *The Road to Excellence: The Making of World-Class Research Universities* (Chapter 1). Washington, DC: World Bank.

Anderson, R.D. (2004). Germany and the Humboldtian model. In R.D. Anderson, Ed., *European Universities from the Enlightenment to 1914*. Oxford, UK: Oxford University Press.

ASEE (1955). *Report on the Evaluation of Engineering Education (Grinter Report)*. Washington, DC: American Society for Engineering Education.

ASEE (1968). *Goals of Engineering Education: Final Report of the Goals Committee*. Washington, DC: American Society for Engineering Education.

Berlin (2015). https://www.hu-berlin.de/en/about/humboldt-universitaet-zu-berlin (accessed February 4, 2016).

Bok, D. (2003). *Universities in the Marketplace: The Commercialization of Higher Education*. Princeton, NJ: Princeton University Press.

Bologna (2015). http://www.unibo.it/en/university/who-we-are/our-history/university-from-12th-to-20th-century (accessed February 4, 2016).

Bush, V. (1945). *Science—The Endless Frontier: A Report to the President on a Program for Postwar Scientific Research*. Washington, DC: National Science Foundation.

Cambridge (2015). https://www.cam.ac.uk/about-the-university/history/early-records (accessed February 4, 2016).

Duderstadt, J.J. (2000). *A University for the 21st Century*. Ann Arbor, MI: University of Michigan Press.

Killian, J.R., Jr. (1985). *The Education of a College President: A Memoir*. Cambridge, MA: MIT Press.

Levin, R.C. (2003). *The Work of the University*. New Haven, CT: Yale University Press.

NRC (1985). *Engineering Graduate Education and Research: Engineering Education and Practice in the United States*. Washington, DC: National Academy Press.

OUP (2015). http://global.oup.com/about/oup_history/?cc=us (accessed February 4, 2016).

Oxford (2015). http://www.ox.ac.uk/about/organisation/history (accessed February 4, 2016).

Padua (2015). http://www.unipd.it/en/university/history (accessed February 4, 2016).

Paris (2015). https://www.sorbonne.fr/en/the-sorbonne/history-of-the-sorbonne/la-fondation-de-la-sorbonne-au-moyen-age-par-le-theologien-robert-de-sorbon/ (accessed February 4, 2016).

PSAC (1962). *Meeting Manpower Needs in Science and Technology. Report Number One: Graduate Training in Engineering, Mathematics, and Physical Sciences*. Washington, DC: President's Science Advisory Committee, The White House.

Rouse, W.B. (2014). *A Century of Innovation: From Wooden Sailing Ships to Electric Railways, Computers, Space Travel, and Internet*. Raleigh, NC: Lulu Press.

RPI (2015). http://rpi.edu/about/history.html (accessed February 4, 2016).

Virginia (2015). http://www.virginia.edu/uvatours/shorthistory/ (accessed February 4, 2016).

West Point (2015). http://www.westpoint.edu/wphistory/SitePages/Home.aspx (accessed February 4, 2016).

3

MISSION AND STRUCTURE

What business or businesses are universities in? How do they organize to pursue these businesses? These questions relate to mission and structure. Christenson and Eyring (2011) and DeMillo (2011), among others, address these issues and conclude that universities need to rethink their value propositions and how they deliver this value. Some of their recommendations are elaborated in later chapters.

Many commentators have suggested that higher education is ripe for market disruption, perhaps by online offerings and unbundling of universities offerings by nimble competitors. However, the rules of the game for accreditation, and the requirement of accreditation for students to be eligible for government-backed loans, currently make such unbundling infeasible (Smith, 2011). Such impedances can hinder the pace of transformation. In this chapter, we begin elaborating the nature of what needs to change.

MISSION

Modern universities have long been in the business of education, research, and service. As discussed in Chapter 2, education was the original mission. Research was added in the early nineteenth century and service to society was added by the mid- to late nineteenth century.

Universities as Complex Enterprises: How Academia Works, Why It Works These Ways, and Where the University Enterprise Is Headed, First Edition. William B. Rouse.
© 2016 John Wiley & Sons, Inc. Published 2016 by John Wiley & Sons, Inc.

Table 3.1 provides the mission statements of the five universities where I have served on the faculty, in chronological order. They are all centered on education and research. They are all also outward looking in that they aspire to provide educated people and knowledge gleaned from research to address society's problems and improve the world.

Typical services have included providing leadership in professional societies associated with academic disciplines and extension services such as in agriculture and manufacturing. Services to agricultural were formalized in the United States with the Hatch Act of 1887 that provided grants to land-grant universities to create agricultural experiment stations. This was expanded with the Smith–Lever Act of 1914 that established cooperative extension services connected to land-grant universities.

The State of Georgia founded the Engineering Experiment Station (EES) at Georgia Tech in 1934 to perform applied research of benefit to the State. Early efforts benefitted the textile and forest products industries. Scientific Atlanta, now a major

TABLE 3.1 Mission Statements of Five Universities

University	Mission Statement
Tufts University	"Tufts is a student-centered research university dedicated to the creation and application of knowledge. We are committed to providing transformational experiences for students and faculty in an inclusive and collaborative environment where creative scholars generate bold ideas, innovate in the face of complex challenges and distinguish themselves as active citizens of the world"
University of Illinois	"The University of Illinois will transform lives and serve society by educating, creating knowledge and putting knowledge to work on a large scale and with excellence"
Delft University of Technology	"TU Delft's mission is to make a significant contribution towards a sustainable society for the twenty-first century by conducting groundbreaking scientific and technological research which is acknowledged as world-class, by training scientists and engineers with a genuine commitment to society and by helping to translate knowledge into technological innovations and activity with both economic and social value"
Georgia Institute of Technology	"Technological change is fundamental to the advancement of the human condition. The Georgia Tech community—students, staff, faculty, and alumni—will realize our motto of 'Progress and Service' through effectiveness and innovation in teaching and learning, our research advances, and entrepreneurship in all sectors of society. We will be leaders in improving the human condition in Georgia, the United States, and around the globe"
Stevens Institute of Technology	"To inspire, nurture and educate leaders in tomorrow's technology-centric environment while contributing to the solution of the most challenging problems of our time"

division of Cisco, was formed as a spin-off of EES in 1952. In 1984, EES became the Georgia Tech Research Institute. In 1988, the National Institute of Standards and Technology established the Manufacturing Extension Partnership (MEP) to enhance the competitiveness of US manufacturing. Georgia Tech is the MEP member for Georgia.

Entities like the Georgia Tech Research Institute and spin-offs like Scientific Atlanta can greatly contribute to economic development in regions surrounding universities. MIT is perhaps the best example, as described in a 1997 study (BankBoston, 1997). By that date, almost two decades ago, MIT research had led to the formation of 4000 companies employing 1.1 million employees with sales of $232 billion. This aggregation of companies amounts to the 24th largest economy in the world. This service component of a university's mission is enormous.

Recent decades have seen substantial "mission creep" in many universities. Involvement in economic development has expanded substantially, for example, many universities now have business incubators. Educational programs associated with sustainability, diversity, gender, ethnicity, etc. have grown steadily, with associated increases in administrators and staff. Collegiate sports have become a major industry, as discussed in the following.

These developments and many others have led to the administrative growth discussed in Chapter 5. Some have argued that the resulting institution may resemble a conglomerate more than an institution of learning (de Boer, 2015; Lazerson, 2010). Others have argued the benefits of universities having more efficient business processes (Auguste et al., 2010; Schumpeter, 2011a, 2011b).

Athletics

Where does athletics fit? Collegiate sports, particularly men's basketball and football, generate huge revenues and, if one listens to university presidents, enormous good will among alumni. It is also argued that the profitability of these sports covers the costs of Title IX commitments to women's athletics.

There is much data on this topic and little of it supports these assertions. Desrochers (2013) reviews this data and reaches the following conclusions:

- In 2010, median athletic spending was nearly $92,000 per athlete; academic spending per full-time equivalent student was less than $14,000 for Division IA universities
- Most Division I athletic programs rely on subsidies from their institutions and students. The largest per-athlete subsidies are in those subdivisions with the lowest spending per athlete
- Athletic costs increased at least twice as fast as academic spending on a per capita basis across each of the three Division I subdivisions
- There is little to mixed evidence to support assertions that winning athletic teams lead to better student applicant pools, greater alumni giving (for other than athletics), or regional economic boosts

Thus, only a handful of Division I athletic programs produce a surplus. The vast majority of programs have to be subsidized by academic revenues. In addition, many of the philanthropic gifts from pleased alumni are directed to the support of athletics. If the current debate about compensating college athletes for their services results in compensation, the economics of college athletics will suffer further—at least from the university's perspective.

It is useful to be realistic about the economic impact of athletics. Nevertheless, athletic expenditures are not a major driver of the escalating costs of higher education. This topic will, therefore, receive only cursory further attention throughout this book.

STRUCTURE

As discussed in Chapter 2, the organizational structure typical of universities emerged around 1000 years ago. Figure 3.1, repeated from Chapter 1, provides insights into the structure of universities and suggests insights into alternative configurations.

Ecosystem: Society and Government

The broad enterprise of higher education exists within an ecosystem that embodies an economic model and incentive structure that strongly influence the other levels of the enterprise. The economic model assumes that people will acquire general knowledge

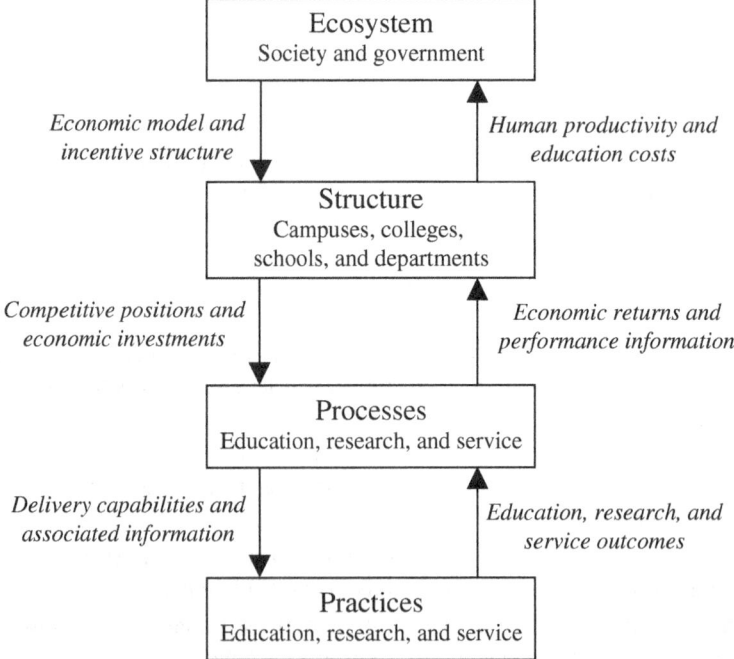

FIGURE 3.1 Multilevel architecture of academic enterprises.

and skills before becoming employed. Employers may provide job-specific knowledge and skills to people after being hired. Thus, individuals have the responsibility to gain general knowledge and skills.

How is this education purchased? K-12 education[1] in the United States is the responsibility of local government. Property tax revenues are typically used to fund local public schools. There are also private K-12 schools where parents usually pay directly for primary and secondary education. The high school graduation rate in the United States increased from 73% in 2005–2006 to 81% in 2012–2013. Graduation rates in major US cities are often in the range of 60%.

High school graduation is typically a prerequisite to matriculate at a college or university. Public universities are usually, but not always, the responsibilities of state government. State legislatures authorize and appropriate budgets to public universities in their state. These budgets effectively subsidize the costs of educating students at these institutions. As discussed in Chapter 6, these subsidies have significantly decreased in recent years.

Private universities receive little, if any, state funding. Thus, students must pay higher prices to attend these institutions. Many private institutions, but by no means all, have substantial endowments. The earnings from these endowments are used, in part, to subsidize the net prices paid by students. Nevertheless, the costs of attending private universities are usually substantially higher than public universities.

The outcomes sought by the ecosystem for investing in higher education are, in general, a healthy, educated, and productive population that is competitive in the global marketplace. This can be characterized as a "public good" in the sense that we all benefit from people being healthy, educated, and productive. This, in principle, enables everyone to contribute to the tax base that provides the resources to support much of the ecosystem.

In recent years, higher education has come to be seen by some as a "private good" that consumers must pay for and should not be subsidized. This sentiment, in part, underlies the recent burgeoning of student debt. Rather than subsidizing students' costs of higher education, they are loaned money that in many cases will require decades to repay. This problem, and some potential remedies, is discussed in Chapter 6.

Structure: Campuses, Colleges, Schools, and Departments

Each university is at the next level of the architecture in Figure 3.1. A university may have multiple campuses, each of which may have significant autonomy. Each campus will have several colleges or schools, each of which will include several departments. Faculty members are associated with departments, which often include multiple academic disciplines and subdisciplines.

This hierarchical organizational structure is a legacy of Bologna, Paris, Oxford, and others. This structure provides a reasonable efficient means for course delivery, assuming that discipline-oriented delivery continues. The implicit assumption is that once students have learned about each puzzle piece, they will know how to make the puzzle. Many academic majors address this need by having capstone "design"

---

[1] Primary education usually refers to grades 1–8, while secondary education refers to grades 9–12, or high school. It is also common for grades 6–9 to be termed junior high school.

courses toward the end of the undergraduate degree. Alumni often report that this capstone course was the highlight of their undergraduate experience.

Organizational structures centered on academic disciplines and subdisciplines do not support large, collaborative research efforts. Problems such as healthcare delivery, sustainable energy, and urban resilience are very much transdisciplinary in the sense that no single or, for that matter, no few disciplines can dominate understanding the problems and exploring alternative contributions to addressing these problems. The overarching limitation of the traditional academic organizational structure is its control of faculty members' promotion and tenure, as discussed in Chapter 7.

Overall, I find the persistence of a structure that was initially designed almost 1000 years ago rather amazing. There must have been some other good ideas in the last millennium. Further, as articulated by architect Louis Sullivan (1896), "Form follows function." The functions of a modern research university in contemporary society are far broader than they were 1000 years ago, and thus the form or structure of universities would seem to need to be broadened or morphed in some ways. This is considered in depth in Chapter 11.

On the other hand, the vested interests and forces aligned with the traditional organizational structure are very strong. The idea of shared governance is deeply ingrained in university cultures. People who are selected for leadership positions are usually firmly committed to the traditional organizational structure. Leadership and governance are discussed in Chapter 4.

Processes: Education, Research, and Service

Organizations make economic investments in delivery capabilities in pursuit of competitive advantage in recruiting the best faculty members and students. These investments are made in the context of the "rules of the game" sanctioned by the ecosystem. These investments are intended to provide economic returns, as well as performance information for deployed capabilities.

There are many types of returns and many time scales over which they are realized. New research laboratories or hires of well-known faculty members can lead to increased student applications with interests in the laboratory's mission or the faculty member's publications. This enables raising admittance standards, leading to better-qualified students, which has been shown to enhance the overall educational experience.

These new laboratories and faculty members should, over time, attract increased research grants from government agencies, foundations, and industry. This will lead to increased publications and eventually citations. As discussed in Chapter 10, this leads to enhanced brand value. This also results in increased student applications at both undergraduate and graduate levels, again over time.

On a longer time scale, increasingly better-qualified students lead to increasingly successful graduates, some of whom eventually become wealthy and want to "give back" to the university. This can be in the form of personal or corporate philanthropy. MIT's success in business formation, as discussed earlier, is a great example of successful students leading to successful companies whose executives then contribute to further the success of their alma mater.

Practices: Education, Research, and Service

The delivery capabilities and associated information provided by the process level of Figure 3.1 enable the practice level to deliver education, research, and services. These practices yield education, research, and service outcomes that include graduate students, research publications, and services to industry, government, and professional societies and associations.

The capabilities drawn upon include classrooms and laboratories but many other processes as well including admissions, registration, bursar, bookstore, dining, dormitories, facilities and grounds, and student health. While this book emphasizes education (Chapter 8) and research (Chapter 9), all of these other services provide value and incur significant costs that contribute to the increasing costs of higher education.

I address these costs by differentiating the costs of administration (Chapter 5) from nonadministrative overhead costs. This allows explicit consideration of the proliferation of administrative head count without detailed accounting of day-to-day operational costs. Admittedly, greater efficiencies in such operations should be an ongoing goal. However, such business process improvements are not central to fundamental transformation of higher education.

Research Centers

There are different types of research centers. A key distinction is the extent to which a center is discipline driven versus problem driven. The former emerges when a group of faculty members and graduate students from the same discipline, for example, supply chain optimization or behavioral economics, join forces to share and attract resources as well as develop a joint brand identity.

The latter type of center is driven by shared interest in a problem domain, for instance, healthcare delivery or urban resilience. Advancing the discipline is very much secondary to the primary goal of contributing to addressing the problems of the domain of interest. This requires an investment strategy focused on the phenomena associated with the problems to be solved.

For large-scale problems like healthcare delivery, it quickly becomes apparent that lack of efficiency and effectiveness cannot be successfully addressed from the perspectives of one discipline. The problems involve medicine, of course, but also economics, public policy, operations research, and behavioral and social science. This realization can help to overcome faculty members' typical penchant for reductionism.

Research centers and institutes that cut across departments, schools, and colleges can provide the support needed to work across disciplines and integrate research products into useful contributions to the larger problems of interest. My first experience of such an organization was the Coordinated Science Laboratory (CSL) at the University of Illinois at Urbana–Champaign. I joined the faculty of the Department of Mechanical and Industrial Engineering (M&IE) in 1974, with a joint appointment in CSL. The core funding of CSL was provided by the Joint Services Electronics Program. I had a small project funded by core funding but most of my funding came from other research sponsors.

Involvement in CSL facilitated my human–machine systems group joining forces with the artificial intelligence group to develop model-based human interfaces to cockpit automation. There was only one project in common, funded by the Air Force, but this common endeavor leveraged the people and resources from ongoing projects with other sponsors. The fact that the two groups were housed on the same floor, with faculty and graduate student offices intermingled, enabled easy interchange of ideas and findings.

The second floor of CSL included a reading room with a large selection of interesting journals and magazines, as well as a large coffee urn that was kept refreshed all day and mailboxes for everyone in the building. The result was that two to three times per day people circulated through this space. This led to many small, sometimes serendipitous, interchanges with faculty from electrical engineering, computer science, information systems, and other disciplines.

I remember chatting over coffee with a colleague from electrical engineering, with whom I seemed to have little in common. I was wrestling with a problem of rapidly generating displays of abstract networks, which human subjects were asked to troubleshoot in terms of finding which network node had failed. The display generation method I had cooked up was painfully slow. I came to the coffee pot in frustration. I mentioned this to him. He said, "Sounds like a circuit routing problem." With that comment, I realized that if I thought about the networks like electrical circuits, I could dramatically simplify my display generation problem—which I did.

My joint appointment with M&IE and CSL was not without problems. I had been attracted to the joint appointment due to my experience at Tufts University as a visiting assistant professor in 1973. I did not enjoy being the only faculty member in my research area. I needed a more robust and energetic intellectual environment.

My 50% appointment in M&IE and 50% in CSL made it difficult to meet expectations of both units. I felt that both units expected 75% of my time and energy. More specifically, there was the constant question of which unit got credit for the overhead monies generated by my grants and contracts—overhead is discussed in Chapter 6. This was very much complicated by the fact that a portion of these monies were returned to the unit credited with generating them. I discuss this complication in more detail later in this chapter.

I spent the 1979–1980 calendar year as a visiting professor at Delft University of Technology in the Netherlands. My appointment was in the Laboratory for Measurement and Control, which included a human–machine systems group. The focus was on process control as well as commercial shipping, which broadened my perspective from aviation.

The nature of my appointment required that I take Dutch classes for five mornings per week. I eventually passed the course, but my Dutch was terrible. This was due, for the most part, to the fact that everyone spoke fluent English. When they heard my awkward Dutch, they immediately switched to English. I got very little practice.

I tried to remedy this by eating in the laboratory cafeteria and joining into various discussions among my Dutch colleagues. This was minimally successful because they would switch to English to help me join in. One discussion group had a profound impact. They were discussing fundamental limits in sports. Building upon biomechanics, work physiology, and other disciplines, they speculated on limits in, for example, running a mile. Could anyone ever run a 3-minute mile? How much weight could anyone be expected to lift?

These conversations got me thinking about fundamental limits in a broad sense. To what extent can we ever really understand human behavior? Can we ever perfectly access people's "mental models"? A set of studies and papers that resulted are summarized in Rouse (2015). The point here is that my experiences in this research center, kibitzing on discussions in the cafeteria, led to ideas and insights that would not have happened if I had stayed in my office and focused on the work at hand.

I returned to Illinois in 1980, now a full professor. I decided that I had served my time on the prairie and wanted to be in a larger urban environment. After interviews at several universities, I accepted an offer from the School of Industrial and Systems Engineering (ISyE) at Georgia Institute of Technology in 1981. Several people moved from Illinois with me—two junior faculty members, one research staff member, and the rest graduate students. We formed the Center for Human–Machine Systems Research and quickly added two more faculty members and numerous graduate students.

Georgia Tech bought the center a Digital VAX 11/780 computer and provided ample space. At this point, I made a decision that I came to greatly regret. United Airlines offered us a DC-8 flight simulator. We had it picked up in Denver and trucked to Atlanta. It was soon installed in our new laboratory space in the newly built ISyE building. We replaced the conventional instrumentation in the cockpit with four large digital display screens that enabled varying the cockpit design for various experimental studies.

The flight simulator was great for laboratory tours but soon became a white elephant. Keeping everything working was very expensive and consumed more and more grant money. Sponsors assumed that the university supported this research infrastructure and we found it difficult to include funds for the simulator in proposal budgets. At the same time, computers and large screen displays were getting less and less expensive. Faculty members and graduate students found this general-purpose equipment easier to use. The simulator was eventually junked, although some of the computational and display resources were retained.

This lesson was not unique. During the 1990s, when I was running the two companies discussed in Chapter 9, I consulted for a couple of university-based research centers. One of the essential challenges faced by these centers was sustaining the research infrastructure they needed. Their host universities were reluctant to provide budgets for such needs and the sponsors expected this to be covered by overhead funds. I will return to the question of how universities support research centers in later chapters.

In 2001, I returned to Georgia Tech to become the chair of ISyE. Based on my experience running two companies for over a decade, I decided to focus my research on enterprises as systems. I was particularly interested in why the many companies who has been my customers found it so difficult to address needs for fundamental change. Thanks to the generosity of Michael Tennenbaum, an alum of Georgia Tech, the Tennenbaum Institute was formed in 2004 to study enterprise transformation. I finished my term as chair of ISyE in 2005 and then spent full time running this research center until 2011.

As Executive Director of Tennenbaum Institute (TI), I reported to the Vice Provost for Research and later the Executive Vice President for Research. This resulted in the overhead monies generated by TI flowing to this office rather than the schools in which faculty members' appointments resided. The chairs of the schools were not happy with this situation. In some situations, this caused faculty members to use TI resources to launch a new endeavor but, once it was successful, they would submit subsequent proposals through their schools. This was completely rational given the incentive and reward system at the university.

In 2010, the Executive Vice President for Research merged 200+ research centers into 10+ integrated centers, creating efficiencies but also much pushback and anger. The same people still headed each of the 200+ entities, but all resources were allocated to the directors of the integrated centers. Gaining and sustaining resources now required convincing another level of management. A significant portion of these resources was used to fund competitive seed grants, usually for junior faculty members.

I retired from Georgia Tech in 2012. Two weeks later, I joined the faculty of Stevens Institute of Technology to found the Center for Complex Systems and Enterprises. Lockheed Martin and Northern Light were charter members, soon joined by Accenture. The research agenda was quite similar to that of the Tennenbaum Institute. The center reports to the Vice Provost for Research. Beyond a laboratory—the *Immersion Lab*—the center has no dedicated space and minimal baseline costs.

As I discuss in Chapter 4, each successive center I have formed has had lower baseline costs. This maximizes the funds for faculty members and graduate students to pursue research topics in the center's portfolio. Faculty salaries are the responsibility of their home schools or departments. The center's funds are mainly used to fund graduate students. The center's brand helps faculty members to secure resources from a range of sponsors.

An alternative model involves faculty appointments in the center. For many years, I have been on the advisory board of such a center at MIT. Having their own faculty has enabled launching center-specific degree programs. Promotion and tenure is handled by the center. Thus, faculty members know that the people evaluating them appreciate interdisciplinary research. The downside is that the center has to compete with traditional departments for faculty slots. Chairs of those departments often feel that a new faculty position gained by the center is a loss for their department. This "zero-sum" game can lead to various dysfunctional behaviors.

Appointments

Of course, for the individual faculty member, there can be questions of where they fit in the structure of the university. As discussed in Chapter 7, faculty members are very well aware of where in the organizational structure their promotion and tenure will be handled. Typically, this happens within the disciplinary departments rather than in interdisciplinary research centers with which they may be affiliated. This can undermine the sustainability of research centers because faculty members want to assure that their disciplinary homes highly value them.

Another possibility is to avoid traditional disciplinary departments. George Mason's School of Engineering started out this way. I served on a Provost's committee that reviewed the School of Engineering's progress after 10 years of operation. Interviews of junior faculty led to frequent comments that they would prefer traditional disciplinary department names rather than multidisciplinary hybrid names. Some said they wanted it to be like it was at the university where they got their Ph.D. Over time, the departments at George Mason have adopted more traditional names.

There is also the possibility of having joint appointments as I had at the University of Illinois. For my last 5 years at Georgia Tech, my appointment was two-thirds in ISyE and one-third in the College of Computing. Between running the Tennenbaum Institute and these two faculty appointments, it was difficult to fulfill people's expectations. I always felt that I did not provide enough value to the College of Computing, although I was always very busy.

CONCLUSIONS

University mission statements typically have much in common with regard to education, research, and service. A substantial problem in recent decades has been mission creep, whereby universities attempt to provide value in varying ways to different constituencies. This inevitably leads to greater costs, which are mostly recouped via tuition increases despite the fact that many of the new activities do not directly benefit those paying tuition.

The multilevel structure of universities means that education, research, and service practices occur in the context of processes that provide capabilities that enable practices. These capabilities are created via investments by departments, schools, colleges, and campuses. These investments are motivated by the "rules of the game" created and sustained by the overall ecosystem.

The traditional structure of universities does not align well with the mission and needs of interdisciplinary research centers. There are inherent conflicts between traditional discipline-oriented departments, schools, and colleges and the crosscutting nature of interdisciplinary research centers. This poses challenges for faculty members who are attracted to larger problems but also want to be successful in securing tenure and promotions.

REFERENCES

Auguste, B.G., Cota, A., Jayaram, K., & Laboissiere, M.C.A. (2010). *Winning by Degrees: The Strategies of Highly Productive Higher-Education Institutions.* New York: McKinsey & Company.

BankBoston (1997). *MIT: The Impact of Innovation.* Boston, MA: BankBoston.

de Boer, F. (2015). Why we should fear University Inc.? Against the corporate taming of the American college. *New York Times Magazine,* September 9.

Christenson, C.M., & Eyring, H.J. (2011). *The Innovative University: Changing the DNA of Higher Education from Inside Out.* San Francisco, CA: Jossey-Bass.

DeMillo, R.A. (2011). *Abelard to Apple: The Fate of American Colleges and Universities.* Cambridge, MA: MIT Press.

Desrochers, D.M. (2013). *Academic Spending versus Athletic Spending: Who Wins?* Washington, DC: American Institutes for Research.

Lazerson, M. (2010). The making of corporate U: How we got there. *Chronicle of Higher Education,* October 17.

Rouse, W.B. (2015). *Modeling and Visualization of Complex Systems and Enterprises: Explorations of Physical, Human, Economic, and Social Phenomena.* Hoboken, NJ: Wiley.

Schumpeter (2011a). How to make college cheaper: Better management would allow American universities to do more with less. The *Economist,* July 7.

Schumpeter (2011b). University challenge: Slim down, focus, and embrace technology. *The Economist,* December 10, 74.

Smith, B. (2011). Disrupting college? Lessons from iTunes. *Clarion Call,* March 22.

Sullivan, L.H. (1896). The tall office building artistically considered. *Lippincott's Magazine,* 57, 403–409.

4

LEADERSHIP AND GOVERNANCE

There is a diverse flow of decisions and requests within the multilevel architecture of universities. The ecosystem level reinforces the values and norms of the overall enterprise while also deciding on policies and allocating resources. Campuses, colleges, schools, and departments also decide on policies and allocate resources, as well as approve hiring, promotion, and tenure of faculty members, based on recommendations emerging from the process and practice levels. As discussed in Chapter 7, there are processes for hiring, promotion, and tenure that are usually carefully followed to avoid subsequent objections. There are also processes for approving curricula and courses, scheduling courses and classrooms, and so on.

All of the above is facilitated via a mixture of leadership and governance. Leadership facilitates creating the vision, choosing directions, and securing and allocating resources. Without a good measure of top-down leadership, typical academic cultures would lead to a fragmented enterprise that tried to be all things to all people. As elaborated in the next section, success depends on leaders *not* becoming just managers. University faculty members can be inspired by good leadership, but alienated by strong management.

Governance is the process whereby decisions get made. Universities typically operate via shared governance between the administration and board of trustees or regents and the faculty. The administration and board govern securing and allocating resources for budgets, facilities, etc. The faculty governs curricula and courses, awarding of degrees, and hiring, promotion, and tenure. The faculty governs in the sense that they make recommendations to the administration and board. Usually, but not always, the recommendations are approved.

Universities as Complex Enterprises: How Academia Works, Why It Works These Ways, and Where the University Enterprise Is Headed, First Edition. William B. Rouse.
© 2016 John Wiley & Sons, Inc. Published 2016 by John Wiley & Sons, Inc.

In this chapter, I address both leaderships and governance. First, the need for leadership is articulated and the nature of good leadership elaborated. Next, faculty governance is discussed. I elaborate various leadership and governance issues with vignettes from my experience. In particular, I expand on experiences leading university research centers.

LEADERSHIP

A few years ago (Rouse, 2011), we conducted two surveys—an initial smaller survey and a subsequent much larger survey—of executives and senior managers about the necessary competencies to successfully lead the transformation of a large enterprise. The competencies they were asked to rate were identified from in-depth analyses of four successful transformations. The 18 competencies identified ranged from leadership to management to technical knowledge and skills.

Results of the surveys showed that respondents were less concerned with more operational competencies such as supply chain management and industrial engineering. In contrast, they placed great emphasis on leadership, vision, planning, strategy, culture and change, and collaboration, teamwork, and social networking. These ratings reflect the need for abilities to articulate the requirement for fundamental change, a vision of the "to-be" enterprise, and a compelling overall strategy and high-level plan for transforming from the "as-is" to the "to-be" enterprise. This is quite consistent with the literature on leadership and change (e.g., George, 2006; Kotter, 1995, 1996).

Leadership was the key to success. Management and technical knowledge and skills were rated much lower, but not because they were unimportant. Interviews of several survey respondents indicated that management and technical knowledge and skills were typically readily available in most enterprises and leadership knowledge and skills were not.

Leadership and Change

All commentators on organizational change agree that leadership plays a central role in success. In my experience, sustained commitment by top management is key to almost every challenge of strategic management. Without committed leadership, energies for change dissipate and the status quo dominates. Another possibility is that change becomes destructive, undermining the organization and perhaps even threatening its existence.

Collins and Porras in *Built to Last* (1994) provide what they call lessons of alignment for CEOs, managers, and entrepreneurs. They argue that a primary role of leaders is to align the organization with desired changes by:

- Painting the whole picture—how everything fits together
- Sweating the small stuff—details are how things get done
- Clustering, rather than shotgunning—coherency is important

- Swimming in your own current, even if it's against the tide—walk the walk
- Obliterating misalignments—consistency is essential
- Keeping universal requirements while inventing new methods

Collins (1999) later elaborated the notion of catalytic mechanisms for enabling or fostering change. He suggests that such mechanisms have several common characteristics:

- Produce desired results in unpredictable ways
- Distribute power for the benefit of the overall system, often to the great discomfort of those who traditionally hold power
- Have a sharp set of teeth
- Attract the right people and eject viruses
- Produce an ongoing effect

Note how both the earlier book and more recent article emphasize the role of leadership beyond coaching, mentoring, and cheerleading. Leaders set carefully thought-out and well-articulated standards and then assure that the organization operates in accordance with these standards. Deviants are "obliterated" or "ejected." This is a far cry from an "anything goes as long as you meet your numbers" mentality. This role of leadership, the authors assert, is key to building enterprises that achieve sustained growth decade after decade.

Day (1999) discusses how to overcome obstacles to becoming value driven, which I have found is the most common form of change sought by enterprises. His six overlapping stages of change emphasize the role of leadership:

- Demonstrating leadership commitment
- Understanding the need for change
- Shaping the vision
- Mobilizing commitment at all levels
- Aligning structures, systems, and incentives
- Reinforcing the change

This guidance is quite consistent with that of Collins and Porras. Indeed, there appears to be little disagreement among various pundits on the roles of leaders in achieving change. This immediately begs the question of why change initiatives are not more often successful.

Pfeffer and Sutton (1999) discuss this phenomenon in terms of a knowing–doing gap. This gap has been the downfall of numerous high-profile top managers. Charan and Colvin (1999) reported that 70% of high-profile chief executives that were fired could not be attributed to a lack of customer-driven vision, strategy, and plans. Instead, these CEOs were fired because they could not get their organizations to embrace the necessary changes and execute the plans for accomplishing them. They had some of the necessary ingredients for change, but they lacked the sufficient ingredients as embodied in the highly rated competencies discussed earlier.

Bridging the knowing–doing gap, they argue, involves transforming knowledge of what needs to change into implementable programs of action. Their guidelines for action include the following:

- Why before how—philosophy is important
- Knowing comes from doing and teaching others how
- Action counts more than elegant plans and concepts
- There is no doing without mistakes—what is the enterprise's response?
- Fear fosters knowing–doing gaps—so drive out fear
- Beware of false analogies—fight the competition, not each other
- Measure what matters and what can help turn knowledge into action
- What leaders do, how they spend their time and how they allocate resources, matters

Leaders not only have to set the course. Grand strategy, no matter how compelling, is not enough. Leaders also have to assure that there are programs of action that appropriately reflect the driving strategies. Leaders also have to assure that these programs of action are implemented. While command and control leadership is no longer the norm, laissez-faire leadership is not the alternative. Sustained commitment and involvement are the keys.

Leadership and Time

In several of my books, for example, Rouse (1994, 1996, 1998, 2001), I conclude by summarizing lessons learned in terms of likely success factors for creating and implementing plans, repositioning enterprises in changing markets, and avoiding organizational delusions. In one of my many seminars on these topics, a participant asked, "I've read several of your books and noted with interest the success factors you have compiled. If you could only recommend paying attention to one factor, across all of the issues you discuss, what one factor would you recommend as the most important factor in success?"

Reflecting on this question—and quickly scanning my mental versions of these lists—the answer was readily apparent. Sustained commitment by an organization's leaders is the most important single factor in success. If leaders do not demonstrably commit themselves and this commitment is not visibly sustained, most best laid plans will go awry.

A key indicator of committed leadership is time. Leaders provide attention to the things of most importance to them. Thus, to the extent that the strategic challenges at hand receive attention from your leaders, you can conclude that these challenges are important to these leaders. Similarly, to the extent that these challenges never seem to make it to the head of the queue, you can reach the opposite conclusion.

These two possibilities beg the question of why leaders would not be concerned with strategic issues. One strong possibility is that they are too busy managing to lead.

It may be that they view management as synonymous with leadership. There are clear linkages between leadership and time.

Kouzes and Posner (1987), in their best-selling book on leadership, suggest five fundamental practices of exemplary leaders:

1. Leaders challenge the process—they search for opportunities to change the status quo, and they experiment and take risks during this search.
2. Leaders inspire a shared vision—they envision the future and enlist others in pursuit of this vision.
3. Leaders enable others to act—they foster collaboration, actively involve others, and strengthen others, making each person feel capable and powerful.
4. Leaders model the way—they create standards of excellence and then set an example for others to follow, planning small wins along the way.
5. Leaders encourage the heart—they recognize contributions that individuals make, celebrate accomplishments, and make everyone feel like heroes.

Relative to time, these practices have two things in common. First, if done well, they are time consuming. Second, they cannot be delegated. This does not mean, however, that they always are done well. One reason is that they tend to fall in the important but nonurgent quadrant of Covey's framework (Covey, 1989). It's difficult to pay full attention to leadership responsibilities when you are spending half your time dealing with the urgent but unimportant demands of management.

Leadership responsibilities are often preempted by management tasks. Maccoby (2000) contrasts management as a function (planning, budgeting, evaluating, and facilitating) with leadership as a relationship (selecting talent, motivating, coaching, and building trust). Management functions are much easier to delegate than leadership relationships. However, as noted earlier, delegation is difficult for many managers. Thus, they spend their time managing rather than leading.

Not surprisingly, finding good leaders is a major issue for most organizations. Leadership gurus Warren Bennis and James O'Toole (2000) discuss this issue in the context of the impacts of choosing the wrong CEO. They provide a set of questions to keep in mind when interviewing CEO candidates:

- Does the candidate lead consistently in a way that inspires followers to trust him?
- Does the candidate hold people accountable for their performance and promises?
- Is the candidate comfortable delegating important tasks to others?
- How much time does the candidate spend developing other leaders?
- How much time does the candidate spend communicating vision, purpose, and values? Do people down the line apply this vision to their day-to-day work?
- How comfortable is the candidate sharing information, resources, praise, and credit?
- Does the candidate energize others?
- Does the candidate consistently demonstrate respect for followers?
- Does the candidate really listen?

Notice how factors such as time, delegation, and empowering others are laced throughout these questions.

Relative to several well-known successes as CEOs, Bennis and O'Toole conclude that they "are great because they demonstrate integrity, provide meaning, generate trust, and communicate values. In doing so, they energize their followers, humanely push people to meet challenging business goals, and all the while develop leadership skills in others. Real leaders, in a phrase, move the human heart" (p. 172).

It is important to consider what leadership experts say specifically about how leaders spend their time. Kouzes and Posner (1987) make the following points:

- Quoting Warren Bennis, they indicate that "routine work drives out non-routine work" (p. 47)
- "It seems that situations and people conspire to make leaders into bureaucrats" (p. 48)
- With regard to how leaders spend their time, they suggest that "Time is the truest test of what the leader really thinks is important" (p. 202)

Peter Senge, the well-known proponent of learning organizations as explicated in *The Fifth Discipline* (1990) and other publications, reflects on how managers allocate their time:

- "Apparently, the 'ready, fire, aim' atmosphere of American corporations has been fully assimilated and internalized by those who live in that atmosphere" (p. 304)
- "The way each of us and each of our close colleagues go about managing our own time will say a good deal about our commitment to learning" (p. 305)

The essential issue concerns how allocations of time reflect upon leaders. Leaders need to understand that their allocations of time reflect their *real* priorities. Some would assert that "How you spend your time is who you are." This assertion may be too strong. However, there are certainly merits in the argument that this view significantly affects followers' perceptions.

Leaders also need to understand that their allocations of time send strong messages to their followers. Regardless of "official" goals, strategies, and plans, how leaders spend their time usually reflects the true organizational agenda. Thus, for example, education, research, and service may be the stated goals, but leaders' behaviors may reflect higher priorities on maintaining control and assuring continuity.

Stewards of the Status Quo

Leadership roles in universities come with rather limited power but enormous influence. University leaders set the tone, which can range from visionary, risk taking, and mentoring to deadly stewards of the status quo. I have found that the former types of leaders can inspire the university community to pursue great achievements in education, research, and service. On the other hand, the latter types of leaders can and will stifle anything that "rocks the boat." It is easy to infer that the board of regents or board of trustees has told the new leader, "Don't screw up!"

Stewards of the status quo include people and organizations who are determined to keep everything as it is—programs, salary structures, pensions, tenure, and so on. They want all stakeholders' entitlements—legally mandated or socially perceived—to remain just as they are. It is not surprising that most stakeholders will support this position.

Getting people and organizations to buy into and support fundamental change often requires great creativity and compromise. Often, this creativity is focused on inspiring a sense of urgency or precipitating a "burning platform." Once people truly believe that change is inevitable, they will usually constructively engage in deliberations on the nature of change and how it should be implemented. A good example of this is the healthcare delivery system in the United States. There is much valuable dialogue going on currently among a wide range of stakeholders. Few people still believe that this system is fine just the way it is.

Perhaps the biggest impediment to change is when the stewards of the status quo are in charge. When the leaders of the organization constrain thinking to business as usual, perhaps on steroids, fundamental change becomes very difficult if not impossible. It is common for such leaders to use the vocabulary of change, for example, new directions, strategic leaps, and enterprise transformation. However, this is just rhetoric. Their real goals are to keep the troops fed, make sure the trains run on time, and avoid rocking the boat.

This form of leadership is most common in enterprises that are shielded from market forces. Government, education and religion, for example, typically attract and recruit these types of leaders. The process of searching for new leaders in these types of enterprises places enormous emphasis on identifying candidates that will not be disruptive. This does not always succeed and occasional change agents will secure major leadership roles. Frequently, their tenure in these positions is relatively brief.

Stewards of the status quo thwart change to such an extent that the roots of change typically emerge outside of the mainstream. People such as Alexander Graham Bell and Thomas Edison, and more recently Steve Jobs, formed new types of businesses and, over time, fundamentally changed our day-to-day lives. The phone business, lighting business, and portable device business were not just business as usual on steroids. These men and, of course, many other men and women invested little energy in preserving the status quo.

In general, change happens when market forces drive it. When forces for change are prevalent, enterprises led by stewards of the status quo suffer, fail, and disappear. Such forces are beginning to emerge in higher education. As forces for change become prevalent, one can expect to see many higher education leaders who are ill prepared to be other than stewards of the status quo. Then, slowly and painfully, change agent leadership will become more the rule than the exception.

Leading Research Centers

University research centers are delicate organizational systems. They bring together faculty, research staff, and graduate students for several reasons. Centers are often formed as a result of a large NIH or NSF grant or because of a large gift or grant from industry or wealthy alumni. So, there is money on the table and researchers are naturally attracted to funding.

Researchers can also be attracted to the research agenda of the center. They like the center's portfolio, and other researchers involved and want to affiliate with the endeavor. This is also true for graduate students, who are often attracted to the research portfolio but also looking for graduate assistantships. Prudence is needed to identify students who have the potential to make real contributions.

I have found that any university will embrace a research center that is totally externally funded and places no demands on the university. Such centers provide resources, at least in terms of overhead, to pay for use of the brand on letterhead and brochures. If all goes well, they hit a homerun; otherwise they quietly fold after the external resources are expended.

Universities tend to have two strategies regarding research centers. For things that they view as mission critical, for example, nanoscience and genomics, they will invest far beyond any possible returns—except bragging rights. Other things have to earn their way onto the agenda, typically by paying returns far in excess of required investments. This excess is used to further fund mission critical areas.

From having founded and developed four research centers at three universities, I have formulated a few rules of thumb. First, determine whether or not you are mission critical. If you are receiving resources in excess of what you generate, chances are you are mission critical. If you are, in effect, paying taxes on the resources you generate, you are a cash cow, at least as long as the cash lasts.

If you are mission critical, your success is assured—the university needs you to succeed and needs to trumpet your success. If you are a cash cow, consider how you will deal with any potential shortfall of funding. This may require scaling back the center's aspirations. Another possibility is to move the whole center to another university. A commercial spin-off might make sense. This might seem like being disloyal, but keep in mind that the loyalty to your center is limited to the university being willing to take the money you have secured.

While you are still a cash cow at your university of origin—where the center was founded—there are several tactics worth considering. First, do your best to avoid taxes. A primary mechanism to achieve this is to secure funds that allow no overhead charges. Since you are not mission critical and, therefore, not getting a share of overhead, why contribute to the pool?

Why would universities accept such stipulations? Quite simply, they cannot walk away from money on the table. Money received in this way makes you less of a cash cow. However, you are not getting a share of the milk—or meat!—so why should your research center contribute? If you are really good at this, you will find the university administration wanting to talk about how they can better support you.

When I was 12 or so, I came home from Charlie Boyd's farm with a pigeon under my arm. I proclaimed to my mother, "Charlie Boyd gave me a pigeon!" My mother responded, "He didn't give you a pigeon, he got rid of a pigeon." If you are directing a research center, especially one newly founded, you need to learn how to identify and avoid pigeons.

Pigeons, in the context of university research centers, are faculty members who are difficult to work with and/or consistently underperform. Deans and department chairs often tend to recommend pigeons to research center leaders. If you manage to transform their attitudes and performance, you have solved a problem for the dean or department head. If not, it is now your problem.

One of the primary objectives of the leader of a research center is brand development. You want the broad community to see your center as a prime time player in the areas of its research and teaching. My experience is that the university will not help you with this. Their marketing and communications staff members are oriented to serving the needs of the president, provost, etc.

Thus, you need to identify the constituencies with whom you want to communicate, develop the messages and associated packaging to communicate with these constituencies, and create the capabilities and opportunities to communicate. You will, of course, be the primary one to deliver these messages. However, getting other faculty members involved with this messaging can contribute enormously to fostering a shared mental model of the center's vision.

As soon as possible, you want to get to the point that you are not writing all proposals and leading all projects. Mentoring faculty members, particularly junior faculty members, is the way to grow these competencies. An important aspect of this is providing them opportunities to present their research to senior audiences from industry and government, not just academia. Speaking skills, as well as writing skills, benefit from frequent opportunities to use them.

Finally, as the leader of a research center, you should invest little time enhancing your resume and much time doing things that improve others' resumes. Your center needs to be the vehicle for personal growth of faculty, staff, and students. The outcomes from your center may include many articles, books, patents, etc., but the primary product of a university research center is the people who employ their knowledge and skills to address the needs of society, typically from a long-term perspective, but nonetheless as contributions to the common good.

Leadership Experiences

To summarize the importance of leadership, I will use several examples. In 2004, Georgia Tech President Wayne Clough helped me prepare slides for a $5 million "ask" to fund a research institute. Once Michael Tennenbaum informally agreed to this proposal, Wayne followed up and closed the deal. The following year, when the Tennenbaum Institute had been launched, I was encountering considerable faculty pushback with regard to the Institute's ambitious research agenda. Georgia Tech Provost Jean-Lou Chameau wisely counseled me that faculty pushback would dissolve once they saw money to support their graduate students. These two examples epitomize for me the essence of involved, committed leadership.

In contrast, here are some bad examples. I worked with a small team assisting the president of a university to develop a strategic plan. I soon realized that the goal was to develop a strategic plan that could not possibly offend anybody. The resulting plan was laced with compelling platitudes. It did not reflect strategic choices to morph beyond the status quo.

A new premium tuition master's degree program was developed with the support of deans from three colleges at a major university. After 2 years of program design and development, the Board of Regents approved the program. Before the actual launch, the three deans asked the new provost to commit to sharing the "profits" from this degree program. He refused to make this commitment. The three deans withdrew their faculty members from the program.

At another university where I served on an advisory board, several faculty members complained to the provost about too many graduate students and large classes. The provost told them to not accept so many students. I commented on the sizable revenue provided by the professional master's degree programs in question. His retort was that the university did not care about money. This disingenuous response completely repressed what could have been an important dialogue.

Another university's provost proposed a set of metrics whereby he could assess the excellence of faculty members. I used these metrics to evaluate several Nobel Prize winners and National Academy members. I showed the group the results—none of these people would be judged to be excellent using the proposed metrics. The provost asked the group to come up with new metrics. The group never met again.

These examples of bad leadership reflect tendencies to avoid issues and dodge leading. Leaders need to help organizations make good strategic choices, including what new things to pursue and old things to discard. Leaders need to exhibit sound judgment rather than sidestep controversy or create metrics that, in effect, eliminate the need for judgment.

Finally, here are a couple of humorous examples. When I joined the faculty at the University of Illinois in 1974, Helmut Korst was head of the Department of Mechanical and Industrial Engineering. He was an eminent scholar of gas dynamics, Austrian by birth, and fairly formal in manner. At a social event in their home, his wife Rudy told a story of Helmut deciding to fix the garbage disposal himself rather than calling a plumber—after all, he was a Ph.D. mechanical engineer. He fixed it and tested it. Garbage disappeared as expected. This was a huge success until Rudy opened the dishwasher and found it full of garbage! These kinds of stories very much humanize leaders.

In spring of 1981, I was wrestling with the decision of whether to leave Illinois for Georgia Tech. I met with the Illinois Vice Provost of Research to discuss various issues. He tried to convince me of the merits of a Champaign–Urbana lifestyle. He told me that many mornings he sat on his back porch with a cup of coffee and looked out at endless fields of soybeans. He could not see a single tree. Having grown up on the seacoast in heavily wooded New England, all I could say was, "I know." This resulted in a bit of laughter and broke the tension of me explaining to this leader why I was very likely leaving. Humor can be a great way to connect.

To conclude, consider the implications of poor leadership and retention of the best employees. Efron (2013) indicates, "Employees don't leave companies, they leave managers." He suggests six reasons including the manager or leader's lack of vision, lack of connection to the big picture, and lack of empathy, as well as, more broadly, the job's lack of effective motivation, no future, and no fun. Leaders who do not lead risk losing their best people.

GOVERNANCE

Universities are governed at multiple levels, which often contributes significantly to the complexity of the enterprise. As discussed in Chapter 3, overall goals, strategies, and plans, as well as financial resources, tend to flow down the hierarchy. Human resources and outcomes of education, research, and service tend to flow up, at least in the sense of reporting of outcomes.

Governing Boards

Public institutions are governed at the highest level by their state legislatures. They determine base budgets and often salaries and benefits. A substantial portion of legislative control may be delegated to a board of regents, or equivalent, for public institutions. Boards of trustees usually play this role for private institutions.

Burka (2011) provides interesting insights into the impact state-level politics can have on governance. He reports on Governor Rick Perry's attempts to control the University of Texas and Texas A&M, via control of the membership of each institution's Board of Regents as well as use of a conservative think tank to develop a plan for reform. Initiatives included assessing the productivity, that is, profit or loss, of each faculty member and publishing a spreadsheet with the results. The next step would be to split the teaching and research budgets, apparently to emphasize the extent to which research loses money. The Governor also advocated having the State of Texas oversee accreditation rather than the usual accrediting agencies.

Administration and Faculty

The administration of the university sets policies, procedures, and budgets. However, governance is not all top down. Faculty members govern curricula, hiring, promotion and tenure, and a wide variety of more bottom-up things. Numerous committees that, in my experience, can be very time consuming accomplish faculty governance.

Eisenberg, Murphy, and Andrews (1998) provide an interesting view of decision making in the search for a university provost. The search process studied was affected by "sunshine laws," which challenged needs to avoid violating confidentiality. Their high-level description of the process is quite typical:

- Organize a search committee
- Chose a committee chair

- Develop criteria—both job and leadership
- Announce the search—people you want will not respond to ads or form letters
- Evaluate the candidates—winnow set to manageable size
- Invite candidates to interview
- Recommend candidates

I have served on many search committee and they are inherently "pickup teams." They bring together people who do not normally work together and usually include several people who have no experience in prior searches. This is due to needs to represent different stakeholder groups.

Eisenberg and colleagues describe three narratives that characterize what happened in the search studied. Narrative 1 followed a rational integration perspective in terms of the aforementioned normative list of steps, with open meetings and voting.

Narrative 2 reflected a differentiation perspective and emphasized politics and negotiated order. This narrative was less well structured and concerned levels of experience and status, all complicated by the search committee chair becoming a candidate. Some committee members thought about resigning when this happened.

Narrative 3 took a fragmentation perspective, emphasizing the ambiguities in the process and how at many points of the process confusion was the rule and rationality the exception. This narrative was laced with faculty cynicism when the search committee chair became a candidate. Subgroups evaluating initial candidates used different selection methods. There were disagreement and ambiguity about the meaning of "academic values."

The interesting thing about these three narratives is that they are representative, in my experience, of searches for provosts, deans, and chairs. They nominally follow narrative 1 but seem to inevitably lead to mixtures on narratives 2 and 3. This is not surprising in the complex social systems at all levels of the enterprise architecture.

Other Players

Finally, there are external advisory boards, accrediting agencies, and peer reviews. External advisory boards vet overall goals, strategies, and plans. Accrediting agencies assess the extent to which the university is achieving its stated goals. Peer reviews evaluate disciplinary units and, as elaborated in Chapter 7, play a central role in evaluation of tenure-track personnel. All in all, the university is a very complicated mix of hierarchy and heterarchy.

Governance Experiences

I served on the faculty of Delft University of Technology during 1979–1980. For several, but far from all, governance issues, voting rights were allocated one-third to faculty, one-third to staff, and one-third to students. Students were quite successful in recruiting staff to their issues, thereby gaining two-thirds majority over faculty members. As I understand, this aspect of governance has since changed.

In 1980–1981, I served as director of the IE division of the Department of Mechanical and Industrial Engineering at the University of Illinois. There were 10 or so faculty members in the division. We would have faculty meetings about once per month. For any controversial issues, I would discuss them in advance with each faculty member. During the meeting, I would present a summary of each faculty member's position on the issue at hand and point out what I thought was the essence of any disagreements.

This process led to quite efficient faculty meetings. However, it encountered strong pushback. One faculty member complained to the head of the department that I was subverting faculty governance. When I met with the head and this faculty member, I argued that this process saved hours of discussion and kept meetings to a reasonable duration. The faculty member said that she very much enjoyed long faculty discussions and debates. She felt this was an important aspect of being a faculty member. The department head asked me to stop trying to be efficient.

Georgia Tech offered me a position in April of 1981. There were several elements of the offer beyond title and salary. These elements included five faculty slots, laboratory space, and funds for remodeling and equipment. Illinois attempted to counter my Georgia Tech offer. Because of the multifaceted nature of the offer, it was necessary for the Committee on Committees to get involved. This significantly lengthened the process. I accepted the Georgia Tech offer before they could reach a conclusion.

In contrast, after my interview at Georgia Tech, Mike Thomas, the school chair, asked me to send him a letter detailing what I would need to say "yes" to an offer. I sent him a five-page single-spaced letter. He called me a few days later, agreed to everything in my letter, and asked when I was moving to Atlanta. Apparently, Georgia Tech did not have a Committee on Committees.

In the mid-1980s, I served on the Graduate Curriculum Committee of the School of Industrial and Systems Engineering at Georgia Tech. The committee was considering core course requirements for the Ph.D. degree. There was substantial disagreement. One senior faculty member suggested we choose one or two math courses because "More math is always good for you." I asked him if he had any empirical evidence to support this assertion. He was completely nonplussed by this question. I elaborated a bit. He became furious and stormed out of the room.

It is quite common for members of faculty committees to make strong assertions for which they have no supporting evidence. I call this the delusion of the ubiquity of expertise. Faculty members who are experts in, for example, materials science will act as if they are also experts in economics and psychology, for instance. It can be difficult to counter the resulting ill-formed assertions without the faculty member feeling insulted.

I rejoined the faculty of Georgia Tech in 2001 as chair of the School of Industrial and Systems Engineering. There were monthly meetings of the 10 school chairs, led by the Dean of Engineering. I discovered that nontenure-track titles (e.g., adjunct professor, visiting professor) were not standardized across the schools. I asked if it would not make sense to do so. However, the chairs had little appetite for doing this. They each liked their independence.

A couple of years later, we were performing third-year reviews for tenure-track faculty. This review is halfway to the tenure decision after 6 years. (I elaborate on the overall process in Chapter 7.) One candidate was a full professor who was 65 years old. In his 3 years at Georgia Tech, he had accumulated superlative student ratings, but had no research publications.

I asked about the origin of this unusual case. The chair of his school responded, saying that the candidate was an architect of international renown, had award-winning buildings around the world, and was rather wealthy. He loved teaching and had no interest in writing research papers. I asked why we were evaluating him using traditional criteria. The response was that this was the only evaluation process available.

I suggested that we create a new faculty model, professor of the practice, which was common in many leading engineering schools. The dean requested that I draft a proposal and bring it back to the committee of school chairs. I did this with inputs from a variety of faculty members. The proposal was endorsed by the committee and then circulated to the deans and chairs of every college and program at the university. The result was 100% support of the proposal. Consequently, a committee was formed to study this idea. They reported back 1 year later, again with 100% support. It was put to a vote by the faculty and support was almost unanimous. This is a great illustration of the typical pace of faculty governance.

I have had frequent involvement in the University of California, particularly the San Diego campus, but also the Berkeley, Los Angeles, and Merced campuses. When serving on an advisory committee at Berkeley in 2005, I was surprised by the extent of faculty governance. Faculty members contribute to discussions of faculty salaries by pay grade, as well as influence promotions through pay grades. The universities where I have been on the faculty do not provide this level of input to faculty members.

I have been involved in an interdisciplinary graduate program at San Diego for the past 10 years. The proposal for this master's degree took 5 years to get approved. Rigid administrative constraints led to several rounds of revisions. For example, the original proposal indicated that the degree would be jointly administered and awarded by the School of Engineering and the School of Business. The proposal made it all the way to Sacramento before it was discovered that the university system does not allow degrees to be administered and awarded in this way.

During 2008–2010 at Georgia Tech, we developed a proposal for a premium tuition master's degree and took this proposal through graduate curriculum committees in all three colleges involved, the university graduate curriculum committee, and the Board of Regents. The approval process required several steps that made no sense for degree being pursued. We were told to fake the content for these steps in order to get beyond them. The managers of the process were quite supportive, but they had no alternative paths we could take. It took 2 years to get final approval.

This degree program was targeted at midcareer people who wanted to gain knowledge and skills to enable them assuming leadership positions in enterprise transformation efforts in their companies. We decided to design all the courses around 4-hour modules. Students would participate in four of these modules all day

Friday and Saturday once per month for 2 years. We came to recognize that there were good reasons to not complete each course within one semester, so we designed to overall curriculum as a dovetailed set of modules.

The Registrar did not like this. Everything had to be linked to semesters. We argued our case for what we felt was a more effective pedagogy. Finally, the folks from the Registrar's office said, "We understand and you are probably right. However, our IT system cannot handle what you want." Legacy IT systems can be quite an impedance to change.

In 2011, I chaired a review committee at Tsinghua University in Beijing. At the end of our review and deliberations, I had to present our findings. I expected to present these findings to the Dean of Engineering or perhaps the provost. Instead, I had to present the findings to the chair of the Communist Party at the university. The governance process at Tsinghua obviously has a few elements I have not encountered in the United States.

I have had limited involvement at universities with unionized faculties. The University of Rhode Island and, more recently, the University of Connecticut are two where I have been involved. When faculty members are unionized, governance is strongly affected. For example, it limits disciplining disruptive faculty members because union representatives have to be involved.

The graduate students at Connecticut are also unionized and represented by the United Auto Workers. The result has been a crisply defined set of tasks that can be assigned to teaching and research assistants as well as, of course, very firmly defined working hours. This presents some difficulties for faculty members advising Ph.D. students in science and engineering. Such students are, in effect, being paid to write their Ph.D. dissertations. Ph.D. advisors are uncomfortable in the employer–employee relationship that has, to an extent, replaced the mentor–mentee relationship.

CONCLUSIONS

This chapter has discussed the role of leadership as enterprises seek to change, including the importance of how leaders spend their time. The negative impacts of stewards of the status quo were elaborated. Experiences of leading research centers were considered. Examples of good and bad leadership were discussed.

Governance was considered in terms of governing boards, administration and faculty, and other players. Numerous vignettes of governance were discussed. Decision making in universities clearly exhibits the characteristics of complex adaptive systems. Many independent agents influence the process, and much time is typically consumed getting to a decision.

REFERENCES

Bennis, W., & O'Toole, J. (2000). Don't hire the wrong CEO. *Harvard Business Review*, May–June, 171–176.

Burka, P. (2011). Old college try. *Texas Monthly*, April.

Charan, R., & Colvin, G. (1999). Why CEOs fail. *Fortune*, June 21, 68–78.

Collins, J. (1999). Turning goals into results: The power of catalytic mechanisms. *Harvard Business Review*, July/August, 71–82.

Collins, J.C., & Porras, J.I. (1994). *Built to Last: Successful Habits of Visionary Companies*. New York: Harper Business.

Covey, S.R. (1989). *The 7 Habits of Highly Effective People*. New York: Free Press.

Day, G.S. (1999). Creating a market-driven organization. *Sloan Management Review, 41* (1, Fall), 11–22.

Efron, L. (2013). Six reasons your best employees quit you. *Forbes*, June 24.

Eisenberg, E.M., Murphy, A., & Andrews, L. (1998). Openness and decision making in the search for a university provost. *Communication Monographs, 65*, 1–23.

George, W. (2006). Transformational leadership. In W.B. Rouse, Ed., *Enterprise Transformation: Understanding and Enabling Fundamental Change (Chapter 4)*. New York: John Wiley.

Kotter, J.P. (1995). Leading change: Why transformation efforts fail. *Harvard Business Review, 73* (2), 59–67.

Kotter, J.P. (1996). *Leading Change*. Boston, MA: Harvard Business School Press.

Kouzes, J.M., & Posner, B.Z. (1987). *The Leadership Challenge: How to Get Extraordinary Things Done in Organizations*. San Francisco, CA: Jossey-Bass.

Maccoby, M. (2000). Understanding the difference between management and leadership. *Research Technology Management, 43* (1), 57–59.

Pfeffer, J., & Sutton, R.I. (1999). Knowing "what" to do is not enough: Turning knowledge into action. *California Management Review, 42* (1, Fall), 83–108.

Rouse, W.B. (1994). *Best Laid Plans*. New York: Prentice-Hall.

Rouse, W.B. (1996). *Start Where You Are: Matching Your Strategy to Your Marketplace*. San Francisco, CA: Jossey-Bass.

Rouse, W.B. (1998). *Don't Jump to Solutions: Thirteen Delusions That Undermine Strategic Thinking*. San Francisco, CA: Jossey-Bass.

Rouse, W.B. (2001). *Essential Challenges of Strategic Management*. New York: Wiley.

Rouse, W.B. (2011). Necessary competencies for transforming an enterprise. *Journal of Enterprise Transformation, 1* (1), 71–92.

Senge, P.M. (1990). *The Fifth Discipline: The Art and Practice of the Learning Organization*. New York: Doubleday/Currency.

5

ADMINISTRATION

The role and nature of the administrative functions within universities have changed. It used to be that members of the faculty would, for a few years, become department chairs, school deans, provosts, and even presidents, only to return to the faculty ranks when their terms as administrators finished. Now, we have career administrators, people who early in their careers pursue the path of chair to dean to provost to president.

What is administration? In some sense, it is synonymous with management in that it directs and resources people to achieve goals. In an academic institution, administration refers to the functions responsible for the maintenance and supervision of the institution. Thus, like management, administration focuses on planning, organizing, leading, and controlling the work of the institution, as well as setting of objectives, coordination and direction, and evaluation of performance.

Some type of administrative structure exists at almost all academic institutions. As noted earlier, faculty members who are actively involved in academic or scholarly work rarely govern this structure. Most senior administrators are academics that have relevant advanced degrees and no longer actively teach or conduct research.

Administration can be contrasted with leadership as discussed in Chapter 4. Leadership involves the exercise of high-level conceptual skills and decisiveness relative to envisioning the enterprise vision and mission, developing high-level goals and associated strategies, inspiring people, and changing culture. Leadership knowledge and skills are seldom the strong suits of administrators.

Universities as Complex Enterprises: How Academia Works, Why It Works These Ways, and Where the University Enterprise Is Headed, First Edition. William B. Rouse.
© 2016 John Wiley & Sons, Inc. Published 2016 by John Wiley & Sons, Inc.

Mills (2012) discusses the corporatization of universities. He notes that in 2003, only two colleges charged more than $40,000 a year for tuition, fees, room, and board. Six years later more than 200 colleges charged that amount. The period between 2003 and 2009 saw the start of the recession. This both decimated endowments and gave tax-starved states reasons to cut back their support for higher education. In these ways, the recession put new pressure on colleges and universities to raise their prices. More recently, however, he argues that it is not just the economic climate in which our colleges and universities find themselves that determines what they charge and how they operate; it is their increasing corporatization.

If corporatization meant only that colleges and universities were finding ways to be less wasteful, Mills suggests it would be a welcome turn of events. However, an altogether different process has emerged that is focused on status, views students as customers, and increasingly relies on top-down administration.

He argues that the most visible sign of the corporatization of higher education lies in the commitment that colleges and universities have made to win the war perpetuated by the kinds of ranking *US News & World Report* now provides. (This ranking scheme, as well as several others, is discussed in Chapter 10.)

Colleges and universities continue to do whatever they can to boost their US News ranking, especially when it comes to whom they admit. For example, the more students a college or university gets to reject, the higher it is ranked on the all-important US News selectivity scale. This strategy has pushed more and more high school students to go to extremes to win admission.

In the eyes of college administrators, students, especially those who are not on scholarship, have become customers who need to feel satisfied with the campus experience bought for them at prices that now top $50,000 per year at many elite schools. Food courts, spa-like athletic facilities, and elaborate performing arts centers are increasingly common on college and university campuses. Many colleges and universities have increased spending for student services that on a percentage basis outpace their increases in academic instruction and financial aid.

Mills reports that we are currently witnessing the rise of the imperial university with campuses around the globe, particularly and ironically in countries with authoritarian regimes willing to invest in a brand-name university. Colleges and universities that do not have a foreign campus worry about getting left behind. This has caused many universities to compromise on traditional values and norms, for example, human rights.

Mills argues that a new, permanent administrative class now dominates higher education. Not surprisingly, those administrators who occupy the highest ranks in our college and university bureaucracies are those who have professionally benefited the most from corporatization. He reports that 36 private college and university presidents, according to the *Chronicle of Higher Education*, fall into the million-dollars-a-year category and many more are close behind.

A still bigger change in how higher education is managed lies in its growing number of administrators in its ranks. As Benjamin Ginsberg (2011a, 2011b) has pointed out, administrators have become a greater presence in colleges and universities, while faculty have been in decline. In the last 40 years the number of full-time

TABLE 5.1 Students, Faculty, and Administrators at Several Universities

Institution	Students	Faculty	Students/ Faculty	Administrators	Students/ Admin
State University of New York	465,000	34,000	13.7	57,000	8.2
University of California	238,000	19,700	12.1	135,900	1.7
University of Texas	51,313	3,018	17.0	21,000	2.4
University of Florida	49,042	5,106	9.6	20,349	2.4
University of Illinois	44,520	2,548	17.5	7,801	5.7

faculty at colleges and universities has grown by 50%—similar to increases in student enrollment—but in this same period the number of administrators has risen by 85% and the number of staffers required to help the administrators has jumped by a whopping 240%. In the 1970s, 67% of faculty were tenured or on a tenure track. Today that figure is down to 30%.

Between 1947 and 1995, administrative costs increased from 9 to 15% of spending, while overall spending increased by 148%, meaning that administrative spending increased by 235%. Between 1975 and 2005, the number of administrators and managers increased by 66% at public institutions and 135% at private institutions. The national average for private colleges is 9 administrators per 100 students. Vanderbilt has 64, Rochester 40 and Johns Hopkins 31, or 1.6, 2.5, and 3.2 students per administrator, respectively. Table 5.1 summarizes these metrics for several major public universities, based on data gleaned from each institution's website.

NUMBER OF ADMINISTRATORS AND COSTS

Table 5.2 portrays a typical administrative hierarchy in a university. The number of administrators within the organizational structure in Table 5.2 is given by Equation 5.1:

$$\text{No. of Administrators} = 10 + \left[2 + (5 \times 8)\right] + (M \times 6) + (M \times 6 \times N) \qquad (5.1)$$

There are M colleges times six administrators per college (not counting departments) and N departments per college times six administrators per department.

Figure 5.1 shows the number of administrators versus number of colleges and number of departments per college. The number ranges from 64 to 712. Nominal salaries for these administrators are summarized in Table 5.3. Figure 5.2 shows the annual costs of administrators using the nominal salaries from Table 5.3. The costs range from $7.5 million to $91 million.

Figures 5.1 and 5.2 do not include administrative staff members, whose ranks have been growing even faster than the number of administrators. "Since 1987, universities have also started or expanded departments devoted to marketing, diversity, disability, sustainability, security, environmental health, recruiting, technology, and fundraising, and added new majors and graduate and athletics programs, satellite campuses, and conference centers" (Marcus, 2014).

TABLE 5.2 Typical University Organization

President
- Executive Assistant
- Scheduler
- Vice President for Finance
- Vice President for Human Resources
- Vice President for Information Technology
- Vice President for Institution Advancement
- Vice President for Marketing and Communications
- Chief Legal Counsel
- Director of Athletics

Provost
- Executive Assistant
- Scheduler
- Vice Provost for Academic Affairs
  - Executive Assistant
  - Associate Vice Provost for A
    - Executive Assistant
  - Associate Vice Provost for B
    - Executive Assistant
  - Associate Vice Provost for C
    - Executive Assistant
- Vice Provost for Research
- Vice Provost for Strategic Initiatives
- Vice Provost for Innovation
- Vice Provost for Diversity

Deans
- Dean, College of X
  - Executive Assistant
  - Senior Associate Dean
  - Associate Dean for Academic Affairs
  - Associate Dean for Research
  - Associate Dean for Operations
- Dean, College of Y
  - Chair, Department of R
    - Executive Assistant
    - Associate Chair for Graduate Studies
    - Associate Chair for Undergraduate Studies
    - Director of Finance
    - Director of Human Resources
  - Chair, Department of S
  - Chair, Department of T
- Dean, College of Z

The equations used to generate Figures 5.1 and 5.2 are elements of an integrated economic model of universities discussed in Chapter 12. Other elements of this model are described in Chapters 6–10. This model is used to explore the implications of the scenarios elaborated in Chapter 11.

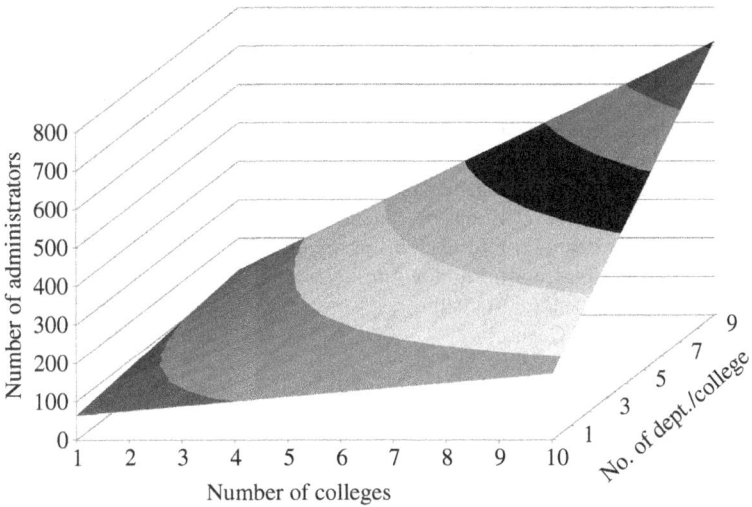

FIGURE 5.1 Number of administrators versus M and N.

PERFORMANCE EVALUATION

A particularly important function of university administration is to evaluate performance. Department chairs evaluate faculty members in their department. Deans evaluate department chairs in their school or college. Provosts evaluate deans. Presidents evaluate provosts. Boards of regents or trustees evaluate presidents.

When I became chair of the School of Industrial and Systems Engineering at Georgia Tech in 2001, I became responsible for evaluating 100 faculty and staff members. Every March, I would spend almost 100% of my time performing evaluations. Fortunately, my predecessor, John Jarvis, had developed a good process that I only slightly modified.

Faculty members were asked to provide an up-to-date copy of their resume, with the accomplishments of the past year highlighted. They were also asked to write a self-review of their accomplishments and plans for the upcoming year. I read each highlighted resume and self-review and completed a one-page form that summarized teacher ratings, publications, proposals, etc. I gave them a rating in each of these areas, as well as an overall rating.

I also copied the key elements of the form into a spreadsheet, which then included the inputs and ratings of all faculty members. The first year, I performed a regression analysis of all this data to see how I was weighting the inputs in making my overall ratings. This showed that I was overweighting one input and underweighting another. To fix this, I did all the reviews over again and checked the weights again, which I then found acceptable.

I provided each faculty member their individual form plus a composite of all the ratings—overall and for teaching, research, and service—organized by faculty rank. I then requested, but did not require, a meeting with each faculty member to discuss his or her evaluation. Most junior faculty members wanted to meet, but senior faculty who were doing well often skipped this opportunity.

TABLE 5.3 Nominal Salaries of Administrators

Office of the President	
President	$500,000
Executive Assistant	100,000
Scheduler	100,000
Vice President for Finance	200,000
Vice President for Human Resources	200,000
Vice President for Information Technology	200,000
Vice President for Institution Advancement	200,000
Vice President for Marketing and Communications	200,000
Chief Legal Counsel	300,000
Director of Athletics	1,000,000
Total	$3,000,000
Office of the Provost	
Provost	$400,000
Executive Assistant	100,000
Scheduler	100,000
Vice Provost for Academic Affairs	300,000
Executive Assistant	100,000
Associate Vice Provost for A	200,000
Executive Assistant	80,000
Associate Vice Provost for B	200,000
Executive Assistant	80,000
Associate Vice Provost for C	200,000
Executive Assistant	80,000
Vice Provost for Research	300,000
Vice Provost for Strategic Initiatives	200,000
Vice Provost for Innovation	200,000
Vice Provost for Diversity	200,000
Total	$2,740,000
Offices of the Deans	
Dean, College of X	$250,000
Executive Assistant	80,000
Senior Associate Dean	175,000
Associate Dean for Acad. Affairs	150,000
Associate Dean for Research	150,000
Associate Dean for Operations	150,000
Dean, College of Y	
Chair, Department of R	200,000
Executive Assistant	80,000
Associate Chair for Graduate Studies	150,000
Associate Chair for Undergraduate Studies	150,000
Director of Finance	100,000
Director of Human Resources	80,000
Dean, College of Z	

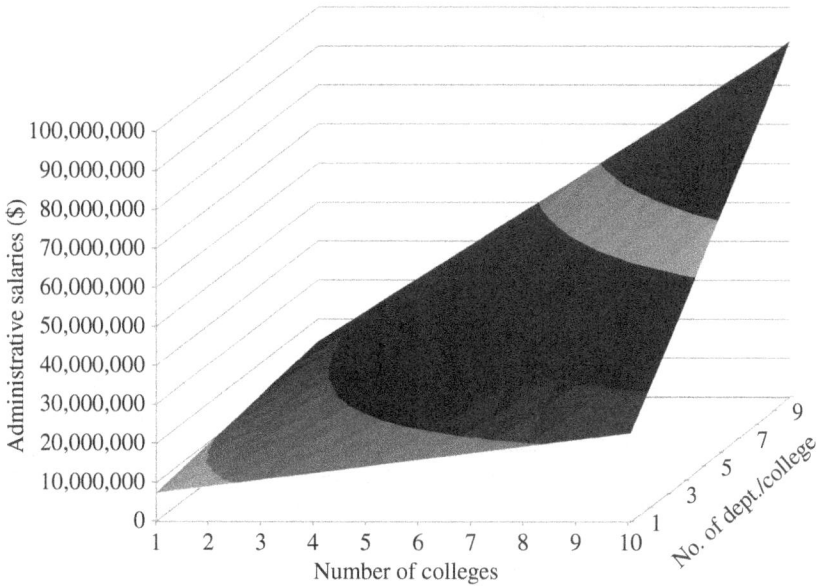

FIGURE 5.2 Costs of administrators versus M and N.

My impression was that most faculty members valued this process, especially because they knew that their ratings had a very significant impact on their raises. I developed the raise plan during the month following the evaluations. I was given a "raise pool" by the dean. Other than complying with the prescribed minimum raise, I was free to allocate raises according to evaluation ratings.

I had an opportunity the following year to apply regression analysis to the salaries of the frontline fundraising professionals on campus, that is, those responsible for raising money, not just counting it. I was surprised to discover that fundraisers' compensation was negatively correlated with funds raised, a statistically significant finding. Looking at the data in more depth, I found that a fundraiser's best strategy was to sign up for a huge goal and not make it. The managers of the fundraising function were quite surprised by these findings.

When I moved to Stevens Institute of Technology in 2012, I learned that they employed a fairly different process. Faculty members complete an online Faculty Annual Review form. This standardized form provides little freedom for anything approaching a self-review. The responses are mostly just numbers, for example, teacher ratings, publications, and proposals. A meeting with your department chair or dean follows completion of the form. I have tried to add a bit of the process from Georgia Tech to the Stevens process, with a smidgeon of success.

The Georgia Tech process is much more time consuming than Stevens' process. However, I prefer it because it provides greater depth of understanding of each faculty member's accomplishments and aspirations. In particular, the composite ratings enabled each faculty member to benchmark themselves versus their peers. This often led to interesting discussions and mentoring opportunities.

CONFLICT MANAGEMENT

One role of administration is managing conflicts. Three classes of conflict are of note:

1. Mission conflicts within organizations
2. Mission conflicts across organizations
3. People conflicts

Conflicts within Organizations

Mission conflicts within organizations typically involve stakeholders disagreeing about where the organization is headed. I encountered this with three initiatives I launched in 2002 shortly after becoming chair of ISyE. The first initiative was the conversion of the school newsletter to a quarterly magazine, *Engineering Enterprise*. The magazine, described later, was launched to position the school as thought leader in the "enterprises as systems" space. Several faculty members resented this initiative because they felt it undermined the school's strong reputation in supply chain and logistics. They wanted to be alone on center stage.

The second initiative was the creation of Georgia Tech Business Network (GTBN) to support young alumni and their entrepreneurial ventures. As discussed in the following, the business school strongly objected to an engineering school using the term business. Many faculty members felt GTBN was a waste of time. We had several events with CEOs of top companies sharing their thoughts on business issues. The venues for these events were packed with young and older alumni. Few faculty members attended.

ISyE had 1200 undergraduates and 500 graduate students. I decided to invite them all to lunch once each semester. Lunch was served buffet style in the triangular courtyard created by our three buildings. Several of the lunches were catered, at no cost, by alumni who owned restaurant chains. Lunch included live music. An enormous number of students, including friends from other schools, joined the lunch. Faculty participation was fair.

These three initiatives cost around $100,000 per year, of a $10,000,000 annual budget. I asked my advisory board to approve this expenditure. They strongly supported the idea of involving our 16,000 alumni and 1700 students. They liked the magazine. Many faculty members would rather have the $100,000 spent to support two graduate students.

Each month at faculty meetings, I provided an update on our budget and expenditures, showing where we were ahead or behind. I quickly learned an important lesson. If I projected a surplus, even just $1000, after the faculty meeting someone would come to my office and ask for this amount for some purpose, typically for travel. From then on, my finance manager and I conspired to always show a modest deficit that would inevitably disappear during the last month of the fiscal year.

Conflicts across Organizations

Conflicts across organizations are typically managed at the level where the two organizations report. An example that I have seen repeatedly when consulting with universities involves overlapping missions between electrical engineering and computer science. Traditionally, electrical engineering had the mandate for the hardware sides of computers, while computer science owned the software side. There were inevitably initiatives where the details of this split were ambiguous. Deans or even provosts had to get involved to resolve such spats.

I mentioned earlier GTBN. The Dean of the Business School strongly objected to this initiative. I suggested that we colead it, but she saw no value in investing time on this initiative. Fortunately, because it was alumni led, the ISyE alums quickly recruited Business School alums to join forces. They eventually built membership to well over 1000 alums.

The idea for GTBN emerged from my thinking that we were devoting all our development resources to courting current wealthy alums but spending little time with alums that might later be wealthy. We invited a group of young alums to a suite we rented at the basketball arena. I asked them, "What can we do to support you, other than sports?" Several responded that they could use some mentoring in their business ventures. GTBN was formed to foster this support.

An interdisciplinary research center where I am on the advisory board was attacked and eventually dismembered by the disciplinary departments that wanted their faculty slots. This university has for three decades or so capped the number of faculty slots, creating a zero-sum game for faculty positions. In other words, any gain by one organizational entity is seen as a loss by other organizational entities. The administration at this university took several years to resolve this.

People Conflicts

People conflicts come in several flavors. There are single-issue conflicts such as the Graduate Curriculum Committee disagreement that I described earlier. These usually work themselves out over time. No intervention is necessary. Sometimes, however, things get more complicated.

A common instance of this involves laboratory space. As discussed in Chapter 6, money and space are the two currencies in academia. Problems arise when faculty members are assigned laboratory space that they no longer actively use. Administrators usually want to reassign this space for other purposes. Faculty can be very creative in how they push back against such changes, in some cases claiming discrimination of one type or another.

There are persistent personal conflicts. I encountered this when I returned to Georgia Tech. Relationships in one research group had deteriorated to such an extent that various members of the group refused to be in the same room together. I talked individually with each of the five members of the group. The levels of disrespect were enormous. Within a few years, all the members of the group had left the school.

Another type of personal conflict I term bad actor conflict. I had several experiences of this while school chair. In one case, an eminent professor went directly to the president with a complaint about the quality of one of his graduate student's desk. The president asked the provost to ask the dean to ask me to explain. Resolution of this simple issue required much more attention than warranted.

Another bad actor example involved another eminent professor who was on the search committee to find my successor. He attempted to get his business partner appointed school chair while not divulging the conflict of interest. Once this was discovered, he was removed from the committee and the candidate in question was disqualified. The provost managed this conflict.

COMPLIANCE AND ABUSE

Universities, especially public universities, have many policies, procedures, and regulations for which the administration has to manage compliance. This can sometimes be taken to extremes as the following vignettes illustrate.

A Georgia Tech faculty member abused the travel reimbursement policy by charging the same travel expenses to two agencies. As a result, the university cancelled all the credits cards of its 6000 employees. Everyone was penalized for one bad actor.

The new policy required employees to have credit cards in their name and front all Georgia Tech expenses, that is, pay the credit card bill first and then get reimbursed. The university determined that many employees were using their cards for personal purchases and personally paying for those purchases.

However, the university had negotiated an $8 reduction of the annual credit card fee for all employees. This amounted to an $8 annual benefit, which meant that the subsidized card could not be used for personal purchases. As a result, the university cancelled all the credits cards of its 6000 employees.

Georgia Tech requires that all requests for reimbursement for food expenditures include an itemized bill so they can assure that no alcohol was consumed. I took three executives to breakfast at Waffle House. The bill for four people was $17. Waffle House does not itemize bills. I was requested to return to Waffle House to get proof that no alcohol was consumed, despite the fact that Waffle House does not serve alcohol.

Some of the aforementioned is due to being a state agency. But such silliness is not limited to the public sector. The Stevens Institute of Technology's compliance function recently began requiring faculty members to prove they have attended conferences, meetings, and functions. So, if you visit a colleague at another university, he or she is supposed to send an email to your supervisor stating that you actually were there. They do not yet require that these communications be notarized!

The reaction to this new policy was rapid and voracious. One dean counseled his faculty to ignore the policy. Upon seeing this email, the compliance function reprimanded the dean for overstepping his authority. One faculty member suggested that all faculty members immediately suspend all travel and stop submitting research proposals that would involve travel if funding was awarded.

MARKETING AND COMMUNICATIONS

Another function of administration is to provide marketing and communications support to colleges, schools, and departments. I have found this problematic at most universities with whom I have been involved. It can be almost impossible for a department or center to get the attention of the marketing and communications function. Thus, you end up doing it yourself.

The aforementioned *Engineering Enterprise* magazine is an example of launching this thought leadership magazine without any help from marketing and communications. From my travels, I learned that the magazine was having the desired impact. The lead interviews of thought leaders from industry, government, and academia were mentioned quite often.

I served on the advisory board at North Carolina State for several years. The department undertook a self-financed (from endowment earnings) marketing and communications plan to increase awareness of program by other departments. This included newsletters, booths at conferences, and other elements. Over just a few years, they were able to significantly improve their *U.S. News & World Report* rankings.

The School of Systems and Enterprises at Stevens took a similar approach. Their marketing and communications plan included publishing a global directory of all programs as well as an electronic newsletter to increase awareness of the program by other schools. This also resulted in significant improvement of their *US News & World Report* rankings.

A significant issue at many universities is the lack of attention to websites and blog platforms. As a result, content gets out of date and updates can take as much as a year. Faculty members have to manage this themselves, which means gaining knowledge and skills far beyond their expertise. I found it more effective to privately host my website and blog because I could not depend on the university.

Book Series

The creation of book series can be another aspect of being seen as thought leaders. At Georgia Tech in 2007, the Tennenbaum Institute launched a book series on enterprise transformation, published by IOS Press in Amsterdam. Five volumes have been published thus far. The volume on healthcare had several thousand advanced orders.

The MIT Engineering Systems Division developed a Series on Engineering Systems, published by MIT Press. The first volume appeared in 2011. Five volumes have been published to date.

The Center for Complex Systems and Enterprises launched the Stevens Institute Series on Complex Systems and Enterprises, published by John Wiley & Sons. The first volume appeared in 2015, with second and third volumes soon to follow.

Such book series help to stake a claim to thought leadership. To the extent that these series attract eminent authors from other universities, associations with such people can also enhance thought leadership. Finally, of course, such series can contribute to codifying and disseminating the best research findings to broader audiences.

THE COSTS OF CONFORMITY

In 2011, I resigned from an administrative leadership position at Georgia Tech, having served for 10 years in that position and an earlier one. The precipitating event involved decisions by senior administrative leadership that I felt limited my abilities to continue in my role. My guess is that it was not intended to have that effect, but the leadership was simply not paying attention to such implications.

However, that was just the straw that broke the camel's back. The nature of "business" processes in academia has long frustrated me. This frustration had been greatly elevated over the previous year with the reorganization of the research enterprise at the university. I agreed with the vision, at least in principle, but the execution was excruciating. Everything moved agonizingly slowly. Almost nothing happened in the summer. Important issues were left hanging for many months, or longer.

In reflecting on this, I realized that effectiveness and efficiency may be the spoken goals, but the real goal is conformity, as well as compliance to assure conformity. We must conform to the policies and procedures of the federal and state governments and those of the university. We must conform to the faculty governance policies and procedures adopted by the faculty senate. This includes operating within the academic culture surrounding this governance model. All this conformity consumes an enormous amount of time and money.

The university's goal is to produce high value outcomes in education, research, and service. That is certainly what students, parents, alumni, sponsors, and the public expect. What about producing these outcomes in a timely and cost-effective manner? It seems to me that the university works as quickly and as cost effectively as it can—within the constraints of all the conformity outlined earlier. The result is that the costs of higher education in the United States are increasing at a faster rate than healthcare, the current poster child for runaway costs.

Raising this issue at several universities with which I am involved has yielded similar, often somewhat arrogant, responses. Higher education sees the value they provide as a given and not to be disputed or even discussed. They see the costs of conformity as inherently justified. Many will say that it is unimaginable that the processes of the university should be redesigned to be more efficient and effective. Most of these processes have been in place for decades and some of them for centuries.

Occasionally, someone in a leadership position decides to consider the redesign of some noncontroversial processes. Typically, this involves forming a committee of faculty and staff of perhaps 15–20 people, given the range of stakeholders that needs to be involved. This committee will meet regularly to discuss and debate at length the nature of the current processes and how they might be improved. All opinions and ideas will be honored and discussed at length.

After a year or two—not counting summers when it is impossible for the committee to meet—a set of recommendations will be created. These recommendations will represent an integration of all ideas discussed, assuring that all committee members can see their ideas in the compilation. The committee will be warmly thanked after they present their recommendations. Perhaps one or two of the ideas on their long list of recommendations will be pursued, as long as they conform to the relevant policies and procedures.

CONCLUSIONS

This chapter has addressed the nature of administration in terms of managing and sustaining a university. The corporatization of academia has resulted in enormous growth of the number of administrators and administrative staff and costs of these personnel.

A relatively simple model was proposed for projecting these costs. This chapter also considered several core administrative functions, including performance evaluation, conflict management, compliance and abuse, and marketing and communications. The costs of conforming to various administrative practices were discussed.

REFERENCES

Ginsberg, B. (2011a). *The Fall of the Faculty: The Rise of the All-Administrative University and Why It Matters.* New York: Oxford University Press.

Ginsberg, B. (2011b). Administrators ate my tuition. *Washington Monthly*, September–October.

Marcus, J. (2014). *New Analysis Shows Problematic Boom in Higher Ed Administrators.* Boston, MA: New England Center for Investigative Reporting.

Mills, N. (2012). The corporatization of higher education. *Dissent*, Fall.

6

MONEY AND SPACE

There are two currencies in academia—money and space. The supply of money can be unlimited if you can convince the world to give it to you. Space, in contrast, is limited at any particular point in time. Consequently, availability and the control of space are often contentious issues on college campuses. This chapter will first address money in terms of overall economics, costs, and revenues, culminating in an outline of an overall economic model of universities. Then, space issues are addressed.

Universities consume enormous resources and are constantly in search of additional resources. Public universities, especially the better ones, used to be able to depend on government support, but legislative support has steadily decreased in recent years—in part, to fund prisons and increasing healthcare costs, for example, Medicaid. Consequently, as discussed earlier, tuitions have soared. The Federal student loan program has enabled students to endure these increases but has led to student loan debt now exceeding all credit card debt.

Universities are constantly on the prowl for Federal research monies although, as discussed in Chapter 9, the hidden costs of pursuing these funds can exceed the level of funding received. Most graduate programs, particularly in science and technology, depend on the availability of F1 student visas, which enable foreign graduate students to often dominate graduate enrollments. Tuition revenues from such students, as well as revenues from executive programs, tend to be among the most profitable and subsidize other academic programs and administrative functions.

Universities as Complex Enterprises: How Academia Works, Why It Works These Ways, and Where the University Enterprise Is Headed, First Edition. William B. Rouse.
© 2016 John Wiley & Sons, Inc. Published 2016 by John Wiley & Sons, Inc.

ECONOMICS OF HIGHER EDUCATION

This section addresses the value of education, economists' views, government subsidies, a potential higher education bubble, and a proposal for public endowment of higher education.

Value of Education

Is education worth the investment? Is it worth it to students and their families who pay tuition? Is it worth it to governments and philanthropists who subsidize such investments in various ways?

We often see publications that report how much greater the lifetime earnings are for someone with a college degree than just a high school diploma. Of course, such assessments are necessarily retrospective. Some have argued that a prospective view will not yield the same premium.

Nelson and Phelps (1966), in a classic paper, discuss investments in humans, technological diffusion, and economic growth. They hypothesize that educated people make good innovators, so that education speeds the process of technological diffusion. In other words, educated people understand new ideas better and adopt them quicker.

They use agriculture as an example, but discuss industry as well, to show a positive relationship between education and adoption of new ideas. They also identify the impact of the economic and social context. Nelson and Phelps conclude, "The rate of return to education is greater the more technologically progressive is the economy."

A few years ago, I integrated the perspectives of various disciplines to address valuation of investments in people's training and education, safety and health, and work productivity (Rouse, 2010). This included reviews of labor economics, human capital economics, defense economics, and engineering economics. Several methods and tools for assessing value were also considered. Various case studies were included.

An overarching conclusion emerged. To the extent that the investing entity is the same entity that realizes the returns on the investment, it is often easy to justify investing in people. When the entities differ, however, the investing entity typically sees the expenditure as a cost, not an investment, and tries to minimize the expenditure.

The Bobby Dodd Institute in Atlanta provides a compelling example of this phenomenon. Bobby Dodd trains mentally challenged people to gain job skills. They place 100% of their trainees in jobs. Their average pay is roughly $20,000. The State of Georgia provides the funding for this training.

Prior to their enrollment in the Bobby Dodd program, these individuals were receiving Social Security disability payments. After successful completion of the program, they receive paychecks rather than Social Security payments. This provides a tremendous return on investment for the Federal government.

However, the State government that made these investments sees none of these returns. Consequently, the State budgets much less money for this program than would be needed to serve all the eligible population. To the State, this budget is a cost, not an investment.

If education is a "private good," then the returns to the students and families paying for education must justify the investments. However, if education is, at least in part, a "public good," then the economic growth created by educated people justifies government investment in subsidizing education. Nelson and Phelps made this point 50 years ago, but it seems that commitment to this principle has waned in recent years.

Economists' Views

Paulsen and Toutkoushian (2008) provide an exposition of how economists look at educational policy issues. They outline the following process, which has some resemblance to the economic modeling thread of this book:

- Identify decision makers
- Identify goals and objectives
- Identify constraints
- Make behavioral and simplifying assumptions
- Develop conceptual model of underlying process
- Determine allocation of resources that maximizes achievement of goals

They apply their approach to the issue of access to higher education. They draw upon human capital theory as a component of labor economics. They recognize the limits of a purely economic approach to this issue and note that educational choices are not just economic choices and one should also pay attention to psychological and sociological components of such decisions.

Winston (1999) addresses the awkward economics of higher education. The exposition begins with the question, "How is higher education different from private sector firms?" He notes that higher education is, typically, a nonprofit enterprise, which cannot distribute surpluses outside of the enterprise. This leads to defining "donative–commercial" nonprofit enterprises. Such enterprises can and do subsidize their customers, selling them a product at a price less that the costs of production.

The subsidies vary depending on where the institution is in the hierarchy of colleges and universities. In 1995, students at public institutions paid 12% of the costs of education; at private institutions they paid 46%. In the past 20 years, however, tuitions at public institutions have risen dramatically and subsidies are much lower.

Winston discusses peer effects in terms of customer-input technology. Students educate both themselves and each other. The quality of education that any student gets depends on the quality of their peers. Peer quality is an input to a university's production function. Thus, in contemporary service science terminology, education is a cocreated service involving faculty, students, and peers.

He argues, "Strong students pay a lower net tuition than weaker ones because they contribute more on the margin to the educational activities of the university and hence get more financial aid." This leads to the suggestion that peer quality can compensate for being taught by graduate assistants and adjuncts.

Winston concludes by outlining four characteristics of the higher education market:

1. All schools sell below cost, subsidizing their customers.
2. Different schools have very different accesses to donative resources.
3. Greater donative resources enable greater quality of peers.
4. Schools' positions in the hierarchy affect abilities to attract quality students.

From a student's perspective, an interesting trade-off emerges. The better the quality of the college, the better the quality of one's peers, which will enhance educational outcomes. On the other hand, the higher the quality of the school, the less likely one will be accepted and the greater the competition once enrolled.

Government Subsidies

When I went to the University of Rhode Island, the cost of my education was highly subsidized by the State of Rhode Island. For my senior year, I was awarded a full scholarship. The amount was under $1000. Times have certainly changed.

What has changed of late is the nature of the subsidy. As State appropriations for higher education have decreased, at least per student, public institutions have raised tuitions. As discussed later in this chapter, these increases have been driven by much more than just declining State appropriations. Nevertheless, students have to pay more.

Students borrow money via Federal loan programs but then must pay these funds back after graduation. McGurn (2011) argues that the availability of government subsidies and loans enables colleges to continually raise tuition. There is no incentive to become more efficient and lower costs.

He also proposes moving beyond standard degree structures. He suggests that subsidies should be tied to students' performance and degree completion time. He thinks that investors might pay for students' education with contracts to receive 10% of their subsequent lifetime incomes. The overall point is that something has to change.

Higher Education Bubble

Wood (2011) reports that Americans have a new willingness to question the value of seeking a traditional college degree. Perceptions have grown that college is no longer a sure bet but a risky proposition. There is also the perception that "A college degree itself doesn't necessarily represent much of anything in the way of intellectual attainment or enhanced practical skill."

He argues that emphasis has shifted from academic rigor to social engagement to attract more students to cover increasing costs, leading to lesser qualified students, easier course requirements, and grade inflation. "On any given day of the week, it is easy to find at least one new way in which higher education sets out to divert itself from the task of teaching undergraduates the knowledge and skills they ought to learn."

Wood blames mission creep at universities (discussed in Chapter 3), capture of the university agenda by interest groups (e.g., for social causes), and people turning political convictions into academic careers. However, the bubble is due to runaway costs, and the bursting of the bubble will come when people cannot afford the lifetime costs of higher education and/or different delivery mechanisms can provide the knowledge, skills, and credentials they need. This is discussed in Chapters 11 and 12.

Public Endowment

Lariviere (2010), president of the University of Oregon, argues for a different way for paying for higher education. He notes that tuition in Oregon has increased an average of 7.5% per year—for the past 38 years! This is far beyond GDP growth.

He proposes that the State promise $60 million per year for 30 years, via an $800 million bond issue, and that private sources match this. The resulting $1.6 billion endowment will yield $64 million the first year and $263 million the 30th year, assuming 9% returns and distributions of 4%. By the 30th year, the endowment will be $6.9 billion.

Endowments are discussed in general in a later section of this chapter. However, one point is particularly relevant now. This mechanism might feasibly pay for higher education. It would ease the student debt problem, but it will not incentivize the efficiencies that universities need to pursue. We need to focus on costs.

COSTS OF HIGHER EDUCATION

Budgeting at universities ranges from amazingly ad hoc to professionally managed. Precedent-based budgeting is common—next year's budget looks like last year's budget with a small increment or decrement depending on how things are going. Program-based budgeting makes a lot of sense but can be difficult to implement when faculty and other resources are shared across programs.

Understanding of costs is key to developing realistic and useful budgets. Prediction and control of costs can be better done if one can directly link costs to programs, functions, and activities. In contrast, if a large portion of costs is aggregated into an undifferentiated overhead pool, abilities to predict and control are undermined.

Cost Disease

William Baumol and William Bowen (1966) and William Baumol (1967) coined the phrase "cost disease" to characterize the lagging productivity growth in service-based industries compared to manufacturing industries. Cost disease is intended to explain the inability of industries such as education, healthcare delivery, city management, etc., to decrease the cost of labor via, in particular, information technology. In healthcare delivery, for instance, providers have found it very difficult to increase the efficiency of their workforce (Rouse & Serban, 2014).

Archibald and Feldman (2006) compare the cost disease theory of Baumol and Bowen to the revenue theory of costs of Howard Bowen (1980) in terms of explaining the increasing costs of higher education. Revenue theory says that costs are matched to revenue available, that is, universities spend more because revenue is available. Their data and analyses support cost disease theory. Thus, universities do not spend more just because they can obtain whatever revenue they want. They spend more because they are unable to influence the labor efficiency of their enterprises.

Cost Analyses

Donna Desrochers and her colleagues at the American Institutes for Research in Washington, DC, have been leading the Delta Cost Project on academic spending. Their 2009 report indicated that average spending on education had increased much more slowly than average tuition. They concluded that tuition increases were not being invested in higher quality except in private research universities (Delta, 2009).

Their 2010 report looked at data for the 2000–2010 decade (Desrochers & Kirshstein, 2010). Their conclusions included the following:

- Community colleges suffered the greatest financial hardships
- Private institutions implemented more widespread cuts than public institutions
- Historic declines of state and local funding could not be recouped by tuition increases
- Private institutions constrained education expenditures even as revenues increased
- Institutional subsidies of students reached a decade-long low
- Colleges and universities did not increase degree productivity

Thus, consistent with cost disease theory, universities did not respond to this turbulent decade by becoming more efficient, with the possible exception of changing their labor mix to include more nontenure track and part-time faculty members.

As discussed in Chapter 3, Desrochers (2013) compared academic spending to athletic spending at NCAA Division 1 universities. She reported that:

- In 2010, median athletic spending was nearly $92,000 per athlete; academic spending per full-time equivalent student was less than $14,000 for Division IA universities
- Most Division I athletic programs rely on subsidies from their institutions and students. The largest per-athlete subsidies are in those subdivisions with the lowest spending per athlete
- Athletic costs increased at least twice as fast as academic spending on a per capita basis across each of the three Division 1 subdivisions
- There is little to mixed evidence to support assertions that winning athletic teams leads to better student applicant pools, greater alumni giving (for other than athletics), or regional economic boosts

Clearly, universities are quite willing to invest substantial amounts in athletic programs with the unwarranted assumption that the by-products of these investments will yield returns that justify such decisions.

Indirect Costs

Ledford (2014) reports on an analysis of indirect costs. These are the costs reimbursed by sponsors of funded research projects. Typically the costs of the time of faculty members, research staff, and graduate students are calculated using the following equation:

$$\text{Total cost} = (\text{Direct salary cost} + \text{Benefit cost}) \times (1 + \text{Indirect cost rate}) \quad (6.1)$$

Indirect costs include a portion of administrators' salaries, facilities costs, and a wide variety of other costs. The indirect cost rate is negotiated with the Federal government. There are substantial differences between negotiated indirect cost rates, ranging from 20 to 85%. Further, the reimbursements actually received may differ substantially from the negotiated rate, for example, negotiated rate of 60% and actual reimbursement of 40%.

Negotiated rates are significantly lower than actual indirect costs. This is due, in part, to administrative costs being capped. This cap emerged in the 1990s after Stanford University admitted to misuse of funds to cover unjustifiable costs, for example, depreciation in the value of the university's yacht. Administrative costs are continually increasing, due to staffing patterns discussed in the following, as well as continually increasing federal compliance requirements.

The implication of this situation is that research funds do not fully cover the cost of research. Further, the "costs of sales" to develop and prepare proposals are not covered at all. In Chapter 9, I show how the probability of a proposal being funded by a federal research agency is steadily decreasing. Hence, the unreimbursed costs of sales are progressively increasing. Consequently, research has to be subsidized by tuition and endowment earnings.

Staffing Patterns

Changing staffing patterns at universities has driven the increase in administrative costs. Desrochers and Kirshstein (2014) provide an analysis of these trends. They report that:

- Professional positions, rather than executive positions, drove the growth of administrative costs (2000–2012)
- Noninstructional student services was the fastest growing salary expense (2002–2012)
- Part-time faculty and graduate assistants account for at least half of the instructional staff

- Part-time faculty and graduate assistants replaced new full-time positions at bachelors and masters institutions
- Average number of faculty and staff per administrator declined by 40% and now averages 2.5 or less (1990–2012)
- The average salary of full-time faculty members has been relatively flat (2002–2010)

Martin (2012) provides similar insights. He reports, "The most striking pattern… is a consistent effort to economize by using lower cost faculty alternatives and to reduce staffing ratios for the nonprofessional staff while the staffing ratios for executive/managerial and professional administrative staff increase by as much as half again from 1987 to 2008." He does not mention cost disease but reaches a similar conclusion, "This trend is particularly remarkable since the overhead trend in the private economy during the same period was in the opposite direction."

Student and Institutional Debt

The revenue from tuition, discussed later, is a cost from students' perspectives. This cost is increasingly covered by student loans. Total student loan debt now exceeds total credit card debt.

Haughwout and colleagues (2015) report on student loan borrowing and repayment trends. Their findings include the following:

- In 2004, 25% of student debt was held by borrowers over age 40; that share climbed to 35% by 2014
- The number of active student loan borrowers peaked in 2010, at about 12 million, and is now down to about 9 million
- The substantial increase in default and delinquency rates in more recent cohorts is almost entirely attributable to borrowers from low- and middle-income areas

Mitchell (2015) reports that the student loan problem is even worse than official figure indicate. In particular, 31.5% of students are at least 1 month behind in payments, compared to 8.5% for auto loans. I have seen projections that many of those with current student loans will still be repaying these loans when their children are in college.

These trends suggest that Wood's (2011) articulation of an emerging higher education bubble is not unreasonable. Chappatta and McDonald (2015) provide another indicator. They discuss Sweet Briar College's decision to close due to its unsustainable financial situation. They reported that Sweet Briar's decision was "Spurring investors to reassess debt from schools facing financial strains similar to those of the 114-year-old women's college in Virginia." Using data compiled by Bloomberg, they indicated that "At least 22 colleges with fewer than 4500 students and rated at best three steps above junk have seen trading in their bonds leap…when Sweet Briar said it would shutter in August." The list included Bridgewater College in Virginia and Mills College in California.

REVENUE: TUITION

For most universities, tuition is the dominant source of revenue. Setting tuition levels is usually driven by overall revenue targets and benchmarking with other universities. In other words, universities charge what they think they can get rather than by bottom-up determination of what education costs.

There is a substantial difference between list prices and what most people actually pay. The average discount can be 40–50%. The reason for this difference was starkly portrayed by a colleague's rather surprising experience at a prestigious private university. He was there with his daughter, and one session was devoted to financial aid.

The director of financial aid explained the process as follows. "Some of you have saved monies to pay our tuition for many years, and you are ready to proceed. Others have hardly saved at all and are worried about how you can possibly pay our tuition. For the first, group, you will pay full tuition which is roughly twice what it will cost to educate your child. The extra revenue this provides will allow us to heavily discount the tuition for the second group. That is how the system works."

What amazing candor! The reason that there is a list price is because a significant portion of people can afford to pay it. Everybody else gets a "scholarship." If that is not sufficient, there are always government loans. The overarching principle is the university sets tuition to achieve revenue goals and uses various mechanisms to assure students and their families can pay it.

Campos (2015) reports that tuition has quadrupled over the past four decades. He observes, "If over the past three decades car prices had gone up as fast as tuition, the average new car would cost more than $80,000." He focuses on where the increased revenue is going.

Considering the costs of faculty, he notes, "Increased spending has not been going into the pockets of the typical professor. Salaries of full-time faculty members are, on average, barely higher than they were in 1970. Moreover, while 45 years ago 78 percent of college and university professors were full time, today half of postsecondary faculty members are lower-paid part-time employees, meaning that the average salaries of the people who do the teaching in American higher education are actually quite a bit lower than they were in 1970."

"A major factor driving increasing costs is the constant expansion of university administration. Administrative positions at colleges and universities grew by 60 percent between 1993 and 2009, which Bloomberg reported was 10 times the rate of growth of tenured faculty positions." For example, "The total number of full-time faculty members in the California State University system grew from 11,614 to 12,019 between 1975 and 2008, the total number of administrators grew from 3800 to 12,183—a 221 percent increase." This growth is the cause of tuition increases, not reduced state funding.

Belkin (2013) discusses how to get college tuition under control, noting that "In the past decade, college tuition has risen three times as fast as the consumer-price index and as fast as medical care." Higher education has replaced healthcare as the poster child for runaway costs.

He argues that there are too many administrative units and too many administrators, and administrative salaries have grown unreasonably. Expanding campus amenities and IT growth, both of which require more administrative staff to manage, has partially driven the growth of administrative staff.

Tuition also subsidizes tenure track faculty members' research pursuits, although faculty salaries, as noted earlier, have not grown. Declining state support is often singled out as the driving force for tuition growth. However, federal subsidies that allow universities to continually raise tuition are what allow tuition to grow, with rapidly expanding student debt as discussed earlier.

Schumpeter (2011) discusses how to make college cheaper. He echoes Belkin in arguing for puncturing administrative bloat and not using tuition to subsidize research. He also proposes consolidating or eliminating programs, increasing class sizes, and shortening degree programs. His overall recommendation is better management, as well as fewer managers.

REVENUE: GOVERNMENT DEPENDENCIES

For the past 150 years, universities have depended on government support, at both state and federal levels. Table 6.1 summarizes key legislation that has provided immense resources to universities, as well as influenced how these resources are used.

Research universities have also played key roles in supporting government efforts. As noted in Chapter 2, former MIT president James Killian (1985) reported that MIT's relationship with the federal government reached new heights with World War II. MIT took on critical challenges, for example, the SAGE missile defense system and the Whirlwind computing project. Faculty members and alumni served in

TABLE 6.1 Key Legislative Dependencies

Year	Legislation
1862	Morrill Act provided federal land to states to be converted to resources to form land-grant universities
1887	Hatch Act provided grants to land-grant universities to create agricultural experiment stations
1890	Second Morrill Act extended opportunity to former confederate states to form land-grant institutions with admissions independent of race
1914	Smith–Lever Act established cooperative extension services connected to land-grant universities
1944	GI Bill signed by President Franklin Roosevelt; by 1947 this resulted in 49% of college students being veterans
1972	Title IX Amendment established that no person shall, on the basis of sex, be excluded from participation or benefits of any education program or activity receiving federal financial assistance
1986	Goldwater–Nichols Act granted universities rights to intellectual property created via federally funded research

TABLE 6.2 Founding of Research Funding Agencies

Year	Founded/Renamed
1794	Marine Hospital Service (MHS)
1887	Hygienic Laboratory
1902	MHS renamed Public Health and Marine Hospital Service (PHMHS)
1912	PHMHA renamed Public Health Service (PHS)
1930	Hygienic Laboratory renamed National Institutes of Health (NIH)
1989	Agency for Health Care Policy and Research (AHCPR)
1999	AHCPR renamed Agency for Healthcare Research and Quality (AHRQ)
1820	Army research conducted at Arsenals
1915	National Advisory Committee for Aeronautics (NACA)
1923	Naval Research Laboratory (NRL)
1946	Office of Naval Research (ONR)
1949	Army Research and Development Division
1948	Office of Air Research
1950	Air Research and Development Command
1950	National Science Foundation (NSF)
1951	Air Force Office of Scientific Research (AFOSR)
1958	NACA renamed National Aeronautics and Space Administration (NASA)
1958	Advanced Research Projects Agency (ARPA)
1958	Federal Aviation Administration (FAA)
1977	Department of Energy (DOE)
1992	Merger of all Army labs into Army Research Laboratory (ARL)
1997	Merger of all Air Force labs into Air Force Research Laboratory (AFRL)
2002	Department of Homeland Security (DHS)

important advisory roles in the federal government. Faculty members, including two MIT presidents, served in senior executive positions, on leave from MIT. Both World War II and NASA's space program resulted in MIT becoming and remaining a national resource, very much a key player in "big science." In the process, MIT was transformed from a technological institute into a powerful university.

Table 6.2 summarizes the founding of various government research funding agencies. The federal government currently provides roughly $40 billion annually to universities. The top 10 recipients received approximately 20% of this total. As discussed in Chapter 9, the competition for these funds is fierce.

REVENUE: FUNDRAISING

American universities, much more so than in other countries, depend on philanthropic gifts from alumni, friends, foundations, and corporations. Many buildings are built by and named for donors who pay a large portion of the costs of construction but seldom the costs of operations. The total endowment of US universities is roughly $0.5 trillion.

Many faculty stars have endowed "chairs" that provide discretionary resources, ranging from zero to substantial amounts. The higher these amounts, the more active faculty are likely to be in soliciting donations for endowed chairs. If chairs are just honorable without discretionary funds, faculty members tend to remain uninvolved and let professional staff handle the fundraising.

Priorities for raising funds tend to be set by the university president, which often substantially disincentivizes college deans, department chairs, and faculty members. The use of gifts and endowment earnings is also often centrally controlled. Universities with more decentralized fundraising operations tend to be more engaging to a wider range of stakeholders.

Fundraising Experiences

When I returned to Georgia Tech in 2001, they had recently completed a highly successful fundraising campaign. As I recall, the goal had been $375 million, but the result was $750 million. To a great extent, this was due to each of the university's larger schools having their own fundraising personnel and being able to tap into their alumni base. A primary result was a very large number of endowed chairs, several of which I was able to fill with renowned researchers.

My first assignment as school chair was to raise $10 million to refurbish our three-building complex. There was little enthusiasm among alumni for this. It is very difficult to generate philanthropic monies for refurbishing projects and especially building maintenance unless the buildings are historically significant. Ours was a set of drab brick buildings built in the early 1980s.

On the other hand, many alumni were very enthusiastic about the prospects of a new building. This would cost $50 million but was much more exciting than the $10 million project. Despite strong lobbying by alumni, there was little enthusiasm from the Georgia Tech leadership. During the period when our building could have been built, the university built seven new buildings for bioscience and technology. Obviously, we were not central to the university's growth strategy.

The president of Georgia Tech asked me if, on my next trip to California, I could present a proposal to a potential donor that was an alum of my school and I had previously met. I asked if I could also bounce another idea off this alum, namely, the idea of forming a research institute to study enterprises as systems, particularly why enterprises find it so difficult to transform in response to major market and technology changes. The president agreed and, as mentioned in Chapter 4, helped me to polish the presentation on this idea.

The result was a $5 million commitment to naming this institute—the Tennenbaum Institute. The donor was not interested in the proposal that originally motivated this visit. At that time, supply chain management and logistics was the dominant research area in the school. The faculty members working in this area were fairly vocal in complaining that the $5 million was not for them.

During my 4 years as school chair, I played a central role in stewarding a major donor so that, soon after I stepped down, my successor could close on a $30 million gift to name the school. Stewarding potential donors was a significant responsibility, involving many lunches and dinners, as well as quite a few trips. These prospects were almost always quite impressive and interesting people, making this an enjoyable part of the job.

In 2005, I served on a visiting committee at Berkeley. When the topic of fundraising came up, the department chair said that he was not allowed to talk with wealthy alums of his department. The fundraising function stewarded such wealthy alums to promote the priorities of the president and provost, not the departments from which they graduated.

Beginning in 2007, I consulted with a major donor to North Carolina State University. One of my tasks was to interview the faculty members in the department and determine what major new thrusts they had both the competencies and energy to pursue. Not surprisingly, several faculty members wanted the $10M endowment to support what they are already doing. The donor disagreed and a well-thought-out plan eventually emerged.

When I joined the faculty of Stevens Institute of Technology in 2012, I found that the fundraising function was completely centralized and focused almost solely on pursuing the president's priorities. As a consequence, faculty involvement in fundraising was minimal. In contrast, my experience at Georgia Tech was that faculty members could play a major role in securing endowed faculty chairs.

Summary

I have found that fundraising can be quite interesting and enjoyable, as you get to meet pretty amazing people. I have also found that different universities have varying approaches to how they raise funds, as well as how they manage these funds and deploy them to benefit their ecosystem.

For example, Fleischer (2015) reports that private equity fund managers at Yale, Harvard, Stanford, and Texas received more compensation that these endowments generated resources for tuition assistance, fellowships, etc. On the other hand, monies for professorships, research subsidies, and maintenance exceeded fund managers compensation. At Yale, these managers achieved 36% annual growth recently. Thus, endowment earnings were far greater than the typical 4–5% earning payouts provided to cover university costs. Fleischer argues that universities should be stopped from such hoarding and be required to spend 8% of the corpus per year.

As indicated earlier in this chapter, Lariviere (2010) proposed that the State of Oregon promise $60 million per year for 30 years, via an $800 million bond issue, and that private sources match this. The resulting $1.6 billion endowment will yield $64 million the first year and $263 million the 30th year, assuming 9% returns and distributions of 4%. By the 30th year, the endowment will be $6.9 billion. This mechanism might feasibility pay for higher education in Oregon, but it will not contribute to curing cost disease.

LESSONS LEARNED

My first encounter with money issues that went beyond funding my research group's activities was in the late 1970s at the University of Illinois. Budget difficulties prompted a faculty meeting to brainstorm about how to attract research monies that would pay for undergrad instruction. The idea was for faculty members to charge a portion of their salaries to research projects but not reduce their teaching loads accordingly. Had this worked, it would not have been sustainable nor have passed any eventual audit.

I moved my research group to Georgia Tech in 1981. By that time, I had developed the habit of projecting my research expenditures for the coming months. This was an easy spreadsheet to develop with a handful of faculty and graduate student salaries and a bit of planned travel. However, my projections consistently disagreed with the monthly financial reports that I received. I discussed this inconsistency with the school's finance director.

She asked me how I spotted the inconsistency. I showed her my spreadsheet lined up versus the reports she sent me. She responded, "Ah. That's your problem. You shouldn't believe the numbers I give you. I certainly don't." I asked her how other faculty members managed their finances. She responded, "They don't."

When I returned to Georgia Tech in 2001, I faced a financial crisis immediately. We received a call from the Board of Regents requesting an explanation for why we were spending our entire annual budget on a single purchase. A bit of sleuthing led us to uncover a purchase order for 10,000 units of an item rather than the intended single unit. It was corrected immediately. What was amazing was that this purchase order made it through the school, college, and university without anyone noticing the discrepancy.

At the same time, we were wrestling with the implementation of a new financial system. It was awful. Nobody trusted it. As a result, each school, independently, created their own spreadsheet-based accounting system. I suggested that we all use the same "shadow" system, but people would only agree if we adopted their system. They did not want to change. As a result, these independent shadow systems persisted for a few years.

In my job as school chair, I quickly realized that few faculty members understood the economics of the school. I created a one-slide presentation that showed how monies from the Board of Regents got to us, as well as how research funds flowed, including overhead. I also presented a one-page budget summary at each faculty meeting, showing revenues and expenditures. As mentioned earlier, I learned to always project a slight deficit to avoid requests for any surplus.

In 2002 and 2008, I served on a visiting committee at Stanford University. During the 2002 review, I encountered a lack of appreciation for faculty members who generated substantial sponsored research monies but whose research was not highly mathematical. The 2008 review was rather different in that faculty members in general had greater appreciation for faculty members who generate substantial sponsored research monies as these resources compensated for lack of funding for more highly mathematical research.

When I first moved to Stevens, I struggled with inadequate IT systems that made it impossible to know balances and encumbrances of contracts and grant accounts. Thus, I was for a period flying blind financially. More recently, I uncovered an enormous accounting error. This insight led to pushback from accounting, but my appeal to top leadership resulted in the error being corrected.

OVERALL ECONOMIC MODEL

Most recently, I developed the computational model of higher education economics discussed in Chapter 12. This has led to several in-depth discussions with senior leaders. They were not aware of much of the data that I have uncovered to support model development. I have been advocating what some people term "evidence-based decision making."

OVERALL ECONOMIC MODEL

Figure 6.1 summarizes the structure of an economic model of complex academic enterprises. It is impossible to show all the variable flows of this model, but this high-level depiction captures the key flows, which will be explained in more detail in Chapters 7–10.

The inputs represent the variables that can be adjusted on the model dashboard discussed in Chapter 12. This is where a user of the model can tailor it to their situation, as well as explore "what if" questions regarding how the enterprise might be transformed to address the scenarios considered in Chapter 11. The model can be modified at deeper levels as well, which is discussed in Chapter 12.

The outputs also appear on the model dashboard. All calculations, represented between the inputs and outputs, are done internally, that is, on other linked spreadsheets. Most of these output variables are self-explanatory, except perhaps brand value, which is discussed in detail in Chapter 10.

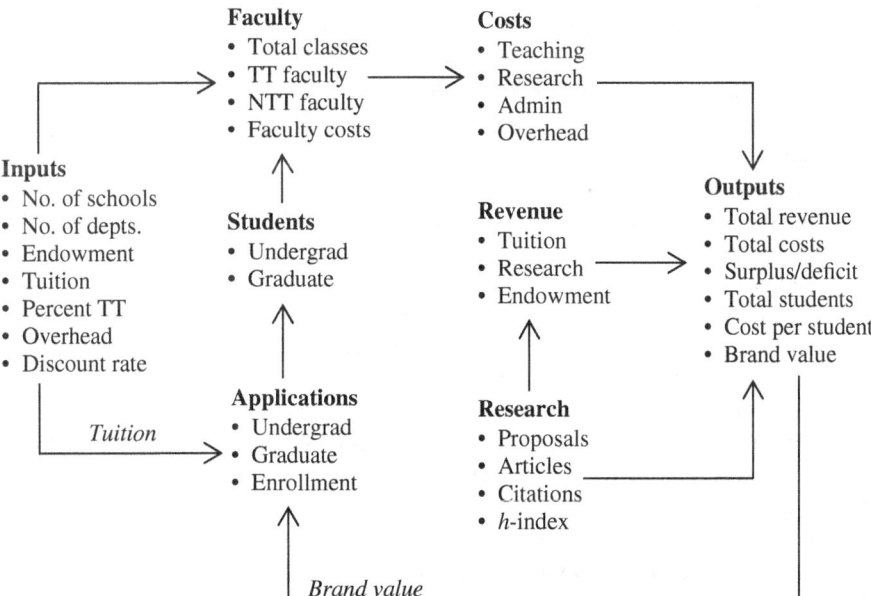

FIGURE 6.1 Overall structure of economic model of academic enterprises.

SPACE

Space has been a major issue on every campus where I have worked. This is due, in part, to the simple fact that the time it takes to create new facilities is much longer than the time it takes to hire faculty members who will utilize the space. In other words, you can increase staff much faster than you can increase space.

The other factor underlying this issue is the great difficulty that universities have in repurposing space. Once space is assigned to particular activities, it can be difficult to reclaim the space once the activities cease. This is particularly true for research space.

Long-defunct research centers can retain space and administrative staffing due to the unwillingness of the leadership to confront faculty members who are no longer productive. When I was a school chair at Georgia Tech, I had great difficulty getting faculty members to give up unused space. In two cases, faculty members pleaded gender and racial discrimination when asked to vacate unused space.

Some universities have the equivalent of "space police," who regularly monitor and assess the extent to which space is being productively employed. An example of this monitoring function involved faculty members lobbying me for research space in a newly rehabilitated building across the street from our main buildings. The space in this large two-story building was assigned to various research groups. With one exception, the space was never used, the space police determined this, and we lost the use of the building.

This tendency to hoard space is due to it being difficult to get assigned space and faculty members' optimism that the big grant is always right around the corner. I really appreciate such optimism, but there comes a time, often after several years, when everyone needs to accept that hoarding no longer makes sense.

For research centers, the cost of space is usually paid from the overhead costs charged to the research sponsors. When the Tennenbaum Institute finally secured its own space a couple of years after it was founded, the Georgia Tech CFO required the Institute to pay for this space despite the fact that these costs were already included in the overhead rate charged sponsors. It took me a couple of years to convince this function that such double charging would never past audit.

The Georgia Tech CFO also required the Institute to pay for refurbishing the space and purchasing the office furniture. When the Institute was required to merge into a larger interdisciplinary institute and consequently forced to move, the previously purchased furniture was "gifted" to the new residents of this space. We were asked to buy new furniture, but my protests eventually caused this requirement to be rescinded.

My recent experiences at Stevens illustrate a different phenomenon. The *Immersion Lab* within the Center for Complex Systems and Enterprises (CCSE) occupies a fairly large space, well over 1000 ft^2. This same space is also the Accenture Innovation Center and the Innovation Lab of the Systems Engineering Research Center.

The result is that this space is very highly utilized. Experiments, demonstrations, meetings, and occasional classes fill up the lab's schedule. Visitors enjoy this very hi-tech space with 8' by 20' 180° touch-sensitive displays. I am glad, however, that CCSE controls the scheduling of the use of the space.

Universities tend to avoid investments of internal funds in space and constantly seek external support for facilities and equipment. As a consequence, deferred maintenance plagues many universities. Thus, it is not unusual to encounter some pretty shabby space at universities. I use this as an indicator of the underlying state of the enterprise.

CONCLUSIONS

This chapter has addressed the economics of higher education, including the value of education, the impacts of government subsidies, and the possibilities of a higher education bubble. The costs of higher education were discussed in terms of cost analyses, indirect costs, staffing patterns, and student and institutional debt. Consideration of revenue sources included discussion of tuition, government dependencies, and philanthropic fundraising. An overall economic model of a university was presented. Finally, the challenging issue of space was addressed.

There is a profound overall implication of this chapter. Change can be very difficult when budgets and space are seen as entitlements. Universities will need much more flexibility if they are to successfully respond to the scenarios described in Chapter 11.

REFERENCES

Archibald, R.B., & Feldman, D.H. (2006). *Explaining Increases in Higher Education Costs*. Williamsburg, VA: Department of Economics, College of William and Mary.

Baumol, W.J. (1967). Macroeconomics of unbalanced growth: The anatomy of urban crises. *American Economic Review, 57*, 415–426.

Baumol, W.J., & Bowen, W.B. (1966). *Performing Arts: The Economic Dilemma*. New York: Twentieth Century Fund.

Belkin, D. (2013). How to get college tuition under control. Wall Street Journal, October 8.

Bowen, H.R. (1980). *The Costs of Higher Education: How Much Do Colleges and Universities Spend per Student and How Much Should They Spend*. San Francisco: Jossey-Bass.

Campos, P.F. (2015). The real reason college tuition costs so much. New York Times, April 4.

Chappatta, B., & McDonald, M. (2015). Sweet Briar's closing plan roils bonds of riskier small colleges. Bloomberg Business, April 19.

Delta (2009). *Trends in College Spending: Where Does the Money Come From? Where Does the Money Go?* Washington, DC: American Institutes for Research.

Desrochers, D.M. (2013). *Academic Spending Versus Athletic Spending: Who Wins?* Washington, DC: American Institutes for Research.

Desrochers, D.M., & Kirshstein, R.J. (2010). *College Spending in a Turbulent Decade: Findings From the Delta Cost Project*. Washington, DC: American Institutes for Research.

Desrochers, D.M., & Kirshstein, R.J. (2014). *Labor Intensive or Labor Expensive? Changing Staffing and Compensation Practices in Higher Education*. Washington, DC: American Institutes for Research.

Fleischer, V. (2015). Stop universities from hoarding money. New York Times, August 19.

Haughwout, A., Lee, D., Scally, J., & van der Klaauw, W. (2015). Student loan borrowing and repayment trends. Liberty Street Economics, April 16.

Killian, J.R., Jr. (1985). *The Education of a College President: A Memoir*. Cambridge, MA: MIT Press.

Lariviere, R. (2010). Saving public universities, starting with my own: The solution is an endowment funded by public and private contributions. Wall Street Journal, November 23.

Ledford, H. (2014). Indirect costs: Keeping the lights on. Nature, November 19.

Martin, R.E. (2012). *Changing Staffing Patterns in Private and Public Research Universities*. Danville, KY: Department of Economics, Centre College.

McGurn, W. (2011). When big government goes to college: The more the Feds try to lower the cost, the worse the problem becomes. Wall Street Journal, April 19.

Mitchell, J. (2015). The student-loan problem is even worse than official figure indicate. Wall Street Journal, April 14.

Nelson, R.R., & Phelps, E.S. (1966). Investment in humans, technological diffusion, and economic growth. *American Economic Review, 56* (2), 69–75.

Paulsen, M.B., & Toutkoushian, R.K. (2008). Economic model and policy analysis in higher education: A diagrammatic exposition. In J.C. Smart, Ed., *Higher Education: Handbook of Theory and Research* (Chapter 1). New York: Springer.

Rouse, W.B. (Ed.). (2010). *The Economics of Human System Integration: Valuation of Investments in People's Training and Education, Safety and Health, and Work Productivity*. Hoboken, NJ: Wiley.

Rouse, W.B., & Serban, N. (2014). *Understanding and Managing the Complexity of Healthcare*. Cambridge, MA: MIT Press.

Schumpeter (2011). How to make college cheaper: Better management would allow American universities to do more with less. The Economist, July 7.

Winston, G.C. (1999). Subsidies, hierarchies, and peers: The awkward economics of higher education. *Journal of Economic Perspectives, 13* (1), 13–36.

Wood, P.W. (2011). The higher education bubble. *Soc, 48*, 208–212.

7

PROMOTION AND TENURE

This topic tends to dominate the thinking of many younger faculty members. The "up or out" nature of tenure creates enormous pressures, although the rapidly declining proportion of tenure track positions across the United States will likely moderate these effects. The basic idea is that, after 6 years usually, but sometimes as much as 10, faculty members are evaluated to determine if the university wants to make a lifetime commitment to them.

The promotion and tenure (P&T) committee reviews the quality of a candidate's publications, the quality of their research funding, the quality of their teaching as assessed by students, and the opinions of external peers on all of the aforementioned. Most P&T committees find it difficult to evaluate candidates in depth mostly due to ever-increasing specialization of academic subdisciplines. Consequently, they rely on the quality of publication outlets and funding sources.

Publication in highly cited journals and funding from the National Science Foundation or National Institutes of Health are deemed to be of high quality because of the difficult wickets one must negotiate to secure such publications and funding. Besides outlining this process, this chapter also considers the pros and cons of the typical P&T system, as well as alternative processes. A model of tenure decision making is discussed later in this chapter.

Universities as Complex Enterprises: How Academia Works, Why It Works These Ways, and Where the University Enterprise Is Headed, First Edition. William B. Rouse.
© 2016 John Wiley & Sons, Inc. Published 2016 by John Wiley & Sons, Inc.

NATURE AND ROLES OF FACULTY

Before delving into the intricacies of promotion and tenure (P&T), we need to discuss the nature and roles of faculty members. This section discusses the nature of academic disciplines, the impact of faculty on education, tenure track (TT) versus nontenure track (NTT) positions, availability of positions, and faculty turnover. These discussions set the stage for consideration of tenure decisions.

Academic Disciplines

Lombardi (2013) characterizes the academic disciplines of faculty members as guilds. The faculty guilds control the agendas and review processes of the basic research funding agencies and the priorities and review processes of the top publication outlets. Through these mechanisms, they control who gets funded and published. Further, their involvement in the P&T process strongly influences whose careers are advanced. Fitting in with these communities and accomplishing what they extol are the absolute keys to success.

Figure 7.1 characterizes these phenomena. The dotted rectangle in Figure 7.1 represents how faculty disciplines both compete and define standards across universities. Members of faculty disciplines at other universities have an enormous impact on P&T processes at any particular university. My sense is that universities may be the only enterprise that allows external parties to largely determine who gets promoted and tenured internally. This has substantial impacts on understanding and modeling the performance of any particular university.

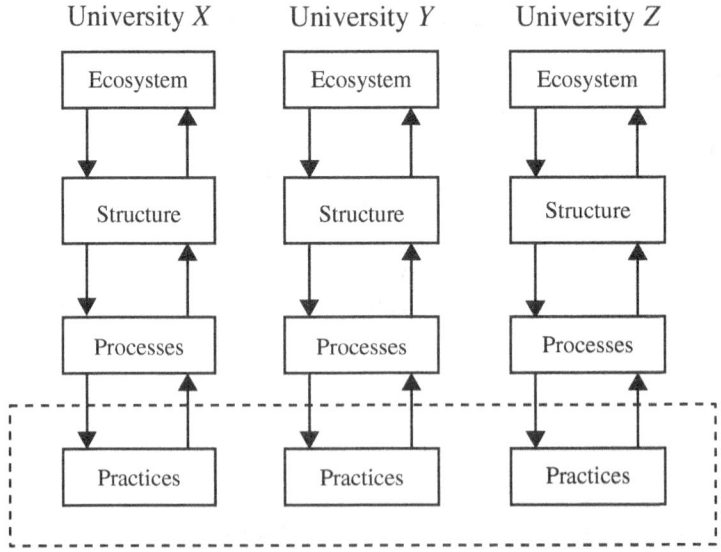

FIGURE 7.1 Hybrid multilevel architecture of academia.

Beecher and Trowler (2001) discuss what they term academic tribes and their territories. Their exposition was motivated by their disagreement with C.P. Snow's *The Two Cultures* (Snow, 1959). Snow argues that society is split into two cultures—the sciences and the humanities. This split, he asserts, is a hindrance to addressing the world's problems.

Beecher and Trowler argue that the following trends have changed the landscape:

- Globalization: Greater vocational orientation
- Massification: Students are less well prepared and teachers are less likely to come from academia; greater vocational orientation
- Altered university–state–industry relations: Increased accountability with emphasis on efficiency and economics. Decreased average teachers' salaries with less job security. More emphasis on transdisciplinary knowledge
- Marketized relationships: Increased power of "customers," increased rivalry among competitors. Commodification of knowledge
- Managerialism within universities: Top-down pursuit of economy, efficiency, and effectiveness
- Substantive disciplinary growth: Growth of knowledge and fragmentation into subdisciplines, along with explosion of journals, for example, 1000 math journals cover 4500 subtopics; 4–8% annual growth in most branches of science

They characterize the impacts on tribes and territories in terms of negativity and resistance to change, perceptions of deprofessionalization, fewer permanent jobs and greater diversity at the periphery, and more internal hierarchies and divisions. Thus, the tribes or guilds fight back by becoming more and more specialized, creating subtribes and subguilds that have ever more esoteric entry requirements.

Faculty Impact

Carrell and West (2010) studied the impact of introductory calculus faculty on achievement in this course and subsequent courses. Less experienced and less qualified professors led to better performance in the introductory course, while more experienced and highly qualified professors, teaching the introductory course, led to better performance in subsequent courses. The former appear to "teach to the test," while the latter appear to provide more depth. Student ratings are higher for teachers who help them to get better grades in the current course—of course, at the time of the ratings, how could students know that their performance in subsequent courses would be better?

Kezar and Maxey (2014) report that the importance of faculty members in the success of students has received much attention. Yet, the point has been reached where over half of all instructional faculty members in nonprofit higher educations are part-time employees. Their much lower pay makes them attractive to universities. This trend is likely to strengthen. Unfortunately, such faculty members typically lack the institutional supports to provide frequent and high-quality interactions with students.

Figlio et al. (2013) ask the question, "Are tenure track professors better teachers?" There provide evidence that NTT faculty members lead to better student performance in the introductory courses they teach, as well as in subsequent courses. An interesting question, which is addressed in later chapters, concerns the extent to which NTT faculty members are able to focus solely on teaching without also having to pursue research and service.

Umbach and Wawrzynski (2005) address the role of college faculty in student learning and engagement. They conclude that "Students report higher levels of engagement and learning at institutions where faculty members use active and collaborative learning techniques, engage students in experiences, emphasize higher-order cognitive activities in the classroom, interact with students, challenge students academically, and value enriching educational experiences." Faculty members at liberal arts colleges are significantly more likely to engage in these behaviors.

TT versus NTT Faculty

Chapter 8 addresses the trade-off between lower-cost NTT faculty and higher-cost TT faculty. NTT faculty members are paid less and typically have double the teaching load of TT faculty. In contrast, as discussed in Chapter 10, TT faculty members' research outputs create the brand value of the university.

Kezar and Maxey (2014), as discussed previously, note that NTT faculty members typically lack the institutional supports to provide frequent and high-quality interactions with students. Fredrickson (2015) reports of NTT faculty members not having offices and, in one instance, holding office hours in their personal automobile.

Fredrickson (2015) notes that tenured or TT faculty members accounted for 80% of all faculty members in 1969. In 2015, 67% of faculty members are NTT; half of those are part time. The impact, he indicates, is that "Thirty-one percent of part-time faculty members are living near or below the federal poverty level. One in four families of part-time faculty are enrolled in at least one public assistance program like food stamps and Medicaid." He reports that many adjuncts are unionizing.

The Economist (2010) discusses the "disposable academic." They assert that doing a Ph.D. is often a waste of time. Ph.D. students and postdocs do much of the teaching and research at universities, but can be stuck in these low-paid positions for many years, and most will never get a TT position. Foreign-born students are much more willing to accept these conditions, which keeps wages down. Overall, the system seems designed to benefit senior faculty and the university in general rather than the students.

Availability of Faculty Positions

There are far more candidates for TT positions than there are positions. NSF (2002) reports that, from 1989 to 1999, there was a 62% growth in the number of nonfaculty, including postdocs, adjuncts, part-timers, and administrators, while the number of full-time faculty grew by just 6%. From 1973 to 1999, the number of white males earning science and engineering Ph.D.s declined from 80 to 40% due to the attractiveness of nonacademic employment. This resulted in an enormous increase in foreign-born Ph.D. students and, subsequently, foreign-born job seekers who have been willing to accept lower-paid postdoc and NTT positions.

Larson et al. (2014) argue that there are either too many Ph.D. graduates or too few academic job openings. They report that an engineering professor in the United States graduates an average of 7.8 Ph.D.s, only one of which can replace him or her. This implies that only 12.8% can attain academic positions. Less than 17% of Ph.D. grads in science, engineering, and health-related fields find TT positions within 3 years. At the same time, faculty retirements are being delayed. They observe that "We have too many PhD STEM graduates for the faculty positions available while there are too few STEM graduates at the BS and MS level to meet the job demands."

Retirements of faculty members could improve the situation at least somewhat. Chronicle (2014) addresses faculty members' attitudes on college retirement and succession planning. They report the following findings:

- 25% of faculty members expect to retire after age 70
- 28% say they are staying longer due to the Great Recession
- 67% say that love of their jobs keeps them from retiring
- 73% of administrators say that faculty staying on beyond normal retirement age limits opportunities for younger faculty members
- 58% are confident of retirement monies but only 28% are confident about healthcare

Administrators would like retired faculty to work as mentors, but faculty members want to stay in their current teaching and research positions. The practice of mandatory retirement ended in 1994 due to expiration of a special provision of the Age Discrimination Act of 1986. Ashenfelter and Card (2002) report that following this change, retirements of 70- and 71-year-old faculty members decreased by two thirds. This, of course, was long before the Great Recession.

Faculty Turnover

Departures of faculty members, for other than retirement, can create opportunities for new faculty members. Of course, if people depart for other university positions, the net number of open positions across academia does not change.

There have been many studies of faculty members' intentions to leave, which have been shown to be highly correlated with actually departures. Interestingly, there are few studies of actual departures. The reason is that departed faculty members tend to not complete and return surveys.

Johnsrud and Rosser (2002) report that "Faculty members are rarely satisfied with their own institutions. They see administrators as incompetent, communications as poor, and their influence declining. This discontent with their institutions is in stark contrast to their satisfaction with their intellectual lives, the courses they teach, and their collegial relationships. Faculty members are dedicated to their work and they love what they do, but they often wonder if they would not be happier doing it somewhere else."

They review extensive previous studies of intentions to leave. A higher salary can pull someone to leave if something about their current position, other than salary, predisposes them to accept the offer. Individuals staying or leaving differed in terms of perceptions of quality of life, time pressure, and chair/department relations. Predictors of intent to leave include frustration due to time constraints and a lack of sense of community.

They introduce a framework with two major dimensions: faculty worklife and morale. The attributes within these dimensions are as follows:

Faculty Worklife
- Professional priorities and rewards
  - Time pressure and lack of personal time
  - Increased oversight and decreased freedom
- Administrative relations and support
  - Confidence in institutional leadership
  - Extent of advocacy for faculty interests
  - System of faculty governance
- Quality of benefits and services
  - Salary and benefits
  - Staffing
  - Working conditions
  - Resource availability

Morale
- Institutional regard
- Mutual loyalty
- Quality of work

Their review of previous studies of morale suggests that high morale could be attributed to distinctive organizational cultures, participatory leadership, sense of organizational momentum, and faculty identification with the institution.

To evaluate their framework, Johnsrud and Rosser studied a 10-campus western university with 2932 faculty members. A 52.1% survey return rate yielded 1511 responses. They found that quality of worklife matters most to the morale of faculty members and the level of morale matters most to the intent to leave. A subsequent study by Rosser (2004) concludes that "Perceptions faculty members have of their worklife have a direct and powerful impact on their satisfaction, and subsequently their intentions to leave."

Daly and Dee (2006) report on faculty turnover intentions in urban public universities. They argue that several challenges affect intent to leave, including the wide range of student preparation levels, life experiences, and learning styles. Other factors include a large population of adult learners, first-generation college students, and English language learners. Expectations of meeting such students' needs conflict with faculty members' research role.

Breiding and O'Meara (2013) report on a survey of University of Maryland faculty members. Their survey resulted in a 47% response rate across Maryland TT faculty members. They found that 4.5% of faculty members would definitely leave the university in the next 2 years; 22.9% were likely to leave. Thirty-nine percent received an outside offer; 44.7% of the offers included salary raises. Full professors were more likely to receive an outside offer than assistant professors; offers to full professors were more likely to include salary increases. Women were significantly less likely to receive an outside offer. 36.1% said the higher salaries were why they would leave; 22.3% said it was due to the offer coming from a more prestigious department of university.

NATURE OF TENURE DECISIONS

Academic tenure grants one lifetime employment at a university. Tenure can be revoked if one very seriously misbehaves, or the university encounters enormous economic difficulties. Such contingencies rarely happen.

I have had the good fortune to have gained tenure at three institutions—first the University of Illinois at Urbana–Champaign, then at the Georgia Institute of Technology, and finally at the Stevens Institute of Technology. I first became a TT faculty member at Illinois in 1974, following a visiting position at Tufts University in 1973. The rules of the game were quite clear:

- Pursue whatever research fascinates you
- Try to get grants or contracts to support this research
- Publish your research in good journals, defined broadly
- Make sure to get above average teacher ratings

That's what I did to secure tenure in 1977. That required just a bit over 3 years. How was that possible? The key was a lack of a fixed timetable for P&T, which also allowed my promotion to full professor by late 1979, effective 1980. So, I was full professor at a top university 10 years after my bachelor's degree. That seldom happens any more. This has very little to do with me and my accomplishments; the academic world has changed.

Now, most Ph.D. graduates first pursue postdoctoral appointments of 1–3 years to create a sufficient publication record to qualify for a TT position. During this time, they are paid roughly twice what a graduate student is paid and half what a TT assistant professor is paid. Someone can easily be living on such meager earnings until their mid-30s. This is part of a trend to delaying tenuring people until well into their 40s. Of course, this is only really a problem in an "up or out" system, that is, either permanent employment or no employment.

During the period of servitude, people strive to attain publications in "A" journals, those that have the highest impact factors in their subspecialty. The impact factor is the average number of times an article in a particular journal is cited, over the past 2 years, by other subsequent articles in any journal. An impact factor of 1.0 is considered mediocre, 2.0 is respectable, and 3.0 is very strong. In fact, all of these numbers are pitiful but not to the cognizati who understand the true significance and impact of one additional citation. Someone, other than your Ph.D. advisor, has read your article and cited it in their article!

Within each subdiscipline, there are usually three or so A journals. Publishing in one of these journals tends to be critical to gaining tenure. Consequently, the number of manuscripts submitted to these journals is enormous, which decreases the probability of having a manuscript accepted for publication (see Chapter 9) and increases the time until it will appear, first online and then in print. This, of course, serves to undermine people's track record when they are considered for tenure.

Another attribute by which faculty in science and technology are judged is the amount of external funding. Not all types of funding are equal. Grant awards from the National Science Foundation (NSF) and National Institutes of Health (NIH) are considered of greater merit because of the peer review systems each of these organizations employs to judge proposals. Typically, one must earn accolades from three to five anonymous members of one's disciplinary or subdisciplinary peer group. The key is to have one's idea fit into the reigning paradigm.

Because these grants count so much toward tenure, they are extremely competitive. As addressed in Chapter 9, the chances of funding at NSF or NIH are steadily decreasing. Further, the awards are often sufficient to only be able to fund one graduate assistant plus one summer month of the principal investigator faculty member. I reviewed an NSF proposal recently that included twelve coprincipal investigators with an annual budget constrained to $300,000. This amounts to $25,000 per investigator; after overhead is deducted, this would cover one summer month of an assistant professor, none of who would likely succeed in the NSF competition.

Other inputs to the tenure process are letters from distinguished members of the candidate's disciplinary or subdisciplinary peer group. These letters need to be glowing, assuring the P&T committee that the candidate is among the few best people in the field and destined for great accomplishments and stardom. The credentials of the letter writers are of significant importance, so their resumes are often part of the package submitted with the letters.

The final part of the candidate's package includes their teacher ratings. If these ratings are above average, then this issue does not receive much attention. Below average ratings, however, usually get considerable attention. Such attention can be avoided by deleting conceptually difficult material from the courses one teaches, making homework and exams very straightforward, and giving high grades to most students. Then, if one can convince students to complete the online evaluation forms, the issue of teacher ratings will be an easy hurdle.

P&T committees usually do not expect candidates to excel in all of the above areas. They do, however, tend to home in on any areas where there are weaknesses, for example, modest funding levels or poor teacher ratings. Strong letters are needed to overcome such weaknesses. It helps tremendously if letter writers come from demonstrably more prestigious institutions and say that the candidate would easily be promoted and/or tenured at their institution.

The result of this overall process is that some people get promoted and tenured and others do not. Those who make it are more likely to be those who conformed to the rules of the game as outlined previously. People who defy the rules have to be very good at what they do. For example, they may publish in other than the A journals but secure enormous grants that pay many people's stipends and salaries. Or, they may be an amazing teacher, publish popular books, and have a local TV show.

Thus, while conformity is the norm, it is not the only path to success. How does creativity and innovation emerge in a culture of conformity? The nonconforming minority is certainly an important source. However, the predominant source is students. They are the "free electrons" in the academic ecosystem. They bind to new ideas and bring the faculty along.

Most, perhaps all, faculty members want to teach and conduct research with the very best students—those that are very talented and highly motivated. Such students often have their own agendas. The intersection of their agendas with faculty members' agendas leads to a creative morphing of agendas and the pursuit of many exciting new ideas.

At this point, tenure comes in very handy. Tenured faculty members can break out of their dominant disciplinary or subdisciplinary paradigms to collaborate with top students who do not recognize and/or accept such constraints. They can look at problems and entertain solutions that do not "fit in." I have done this a few times. Once our results were published, colleagues sometimes asked, "Why on earth did you do that?" A few years later, sometimes longer, some of these ideas were mainstream and readily fundable.

Tenure provides enormous freedom. However, earning tenure can instill a tendency to conform that may later be difficult to escape. Once one gets good at conforming, it can be a pretty comfortable place. One gets accolades, invitations, and eventually awards. The key to creatively moving out of this comfort zone is to engage students in pursuits where, for some aspects of the problems and solutions, they know more than you do. Your expertise and mentoring can take them—and you—to new levels of success.

PROMOTION AND TENURE EXPERIENCES

As noted previously, I have been through the tenure process three times. At the University of Illinois in 1977, I experienced a P&T process that was not driven by a fixed schedule, only by accomplishments. When I joined the faculty of Georgia Tech in 1981, I had to wait 1 year to apply for tenure and never got around to it. When I returned to Georgia Tech in 2001, the rules had changed, and I was granted tenure as school chair and chaired professor. At Stevens in 2012, I was granted tenure as a chaired professor.

The only insight I gained from being a candidate for P&T came from a comment by the department head at Illinois. He mentioned that one of my reference letters indicated that I was one of the top two people in the world in my research area. He said, "Too bad he didn't say you were number one." Later, when serving on P&T committees, I learned that comments like this are common.

What Really Counts

As school chair at Georgia Tech, I served on a committee of school chairs that reviewed all P&T cases in the College of Engineering. This is when I first encountered the great emphasis on NSF and NIH grants. I challenged the dean, who chaired the committee, about such grants being proxy measures of quality when we should be judging quality. I argued that we should not be outsourcing this judgment. Neither he nor any other committee members agreed with me.

At Stevens, I have found that junior faculty are concerned that only NSF and NIH grants count toward P&T. During a recent visit to the University of Connecticut, junior faculty expressed the same concerns. One young faculty member had just received a grant from a company, and he wondered whether he should tell his department head because it might hurt his P&T prospects.

Having served on P&T committees many times over the past 15 years, it has consistently struck me how committee members do their utmost to be fair. No candidate is expected to excel at everything. Granting the benefit of doubt is common. Nevertheless, people who do not measure up do not get promoted and tenured.

Making the Case

Faculty members being considered for P&T are asked to prepare a "tenure package." This package has evolved over the years to become quite elaborate. It used to be that you provided an up-to-date resume and copies of what you felt were your five best journal articles. Now it takes much more preparation.

With the advent of Google Scholar and other services, candidates need to include their number of citations and h-index (discussed in Chapter 9). Often they include a list of all the articles that have cited their articles. Some include a list of the institutions of the authors who cited them.

Typically, the candidate recommends five external people who can serve as references. The P&T committee selects another five. The committee chair, or perhaps the dean or department chair, sends a letter to each of the 10 references asking for their assessment of the candidate. All of the letters received by the deadline plus biosketches of all those providing letters are then added to the package.

This package, now often well over 100 pages in size, along with the recommendation letter of the department P&T committee is sent to the school or college P&T committee. They add their recommendation and send the package to the university P&T committee. Committee members, at all levels, end up with many large files or a lot of paper.

Great care is taken to make sure everyone follows the process, in part, so the committees can defend their decisions if they are later challenged, for example, by an unhappy candidate whose P&T is refused. The first time I served on a P&T committee, I broke one of the rules and my peers reprimanded me.

One of the candidates from my school had written a journal article with two renowned faculty members. The P&T committee members questioned what the candidate had contributed to the paper. After the meeting, I asked the two faculty members what he had contributed. They told me that they had not done any of the work or written any part of the article. The young man had simply added their names, with their permission, because he felt they had given him valuable advice along the way.

At the next meeting of the P&T committee, I reported this finding. I was quickly rebuked for trying to add new information to the process after the deadline. I commented that I was simply correcting an erroneous perception. They said they understood my intent but it was not allowed. The candidate was promoted and tenured.

MODEL OF TENURE DECISION MAKING

To review earlier discussions, P&T decisions are based on numbers of articles published in high-quality journals, numbers of grants funded by favored sponsors, teacher ratings by students, and service to one's profession and the university, all of which is judged by references from faculty members at other universities. My experience is that teacher ratings need to be above average and service needs to be adequate. These metrics, by themselves, will not earn one P&T.

The dominant metrics are numbers of articles published and number of grants funded. Chapter 9 presents historical data for many journals and grant sponsors NSF and NIH. These data are used to fit mathematical models that can be used to project success in submitting articles and proposals. At this point, however, the concern is solely with how numbers of article published and grants funded are used to make P&T decisions.

P&T committees, as well as those writing letters of reference, ask whether they think a candidate will have successful career after being tenured. If they think the likelihood of success is high, they will vote for the candidate. It they think it is low, they will vote against the candidate. Typically the majority rules for the committee, not so for the letters. If a minority of the letters provides a rather negative assessment of the candidate's prospects, this will usually turn the committee negative.

Thus, somewhat simplistically, the committee is trying to separate the low and high probability candidates. If we assume that each article or proposal submitted involves a Bernoulli process with probability of success p, then a series of n submissions can be characterizes by a binomial distribution with a mean of np and variance $np(1-p)$. It would be easy to know n for any candidate, but p is unknown even by the candidate. In Chapter 9, we will estimate p across all people submitting articles or proposals.

It is likely that p varies by individual, but the P&T committee and letter writers cannot make this assessment. Instead, they make a simple low versus high distinction. The binomial distributions shown in Figure 7.2 illustrate this idea for 40 articles or proposals submitted. The committee needs to decide whether a candidate's performance implies a $p=0.2$ or $p=0.5$ process.

Figure 7.3 shows the cumulative probability distributions for the density functions shown in Figure 7.2. These distributions can be used to estimate two important probabilities, the probability of a false acceptance of a $p=0.2$ candidate (PFA) and the probability of a false rejection of a $p=0.5$ candidate (PFR). The committee would

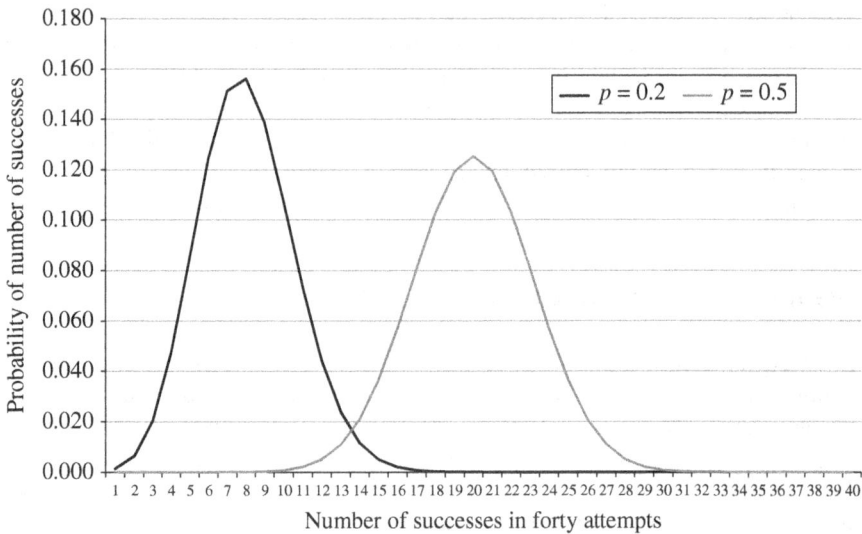

FIGURE 7.2 Binomial probability density functions for $p=0.2$ and $p=0.5$.

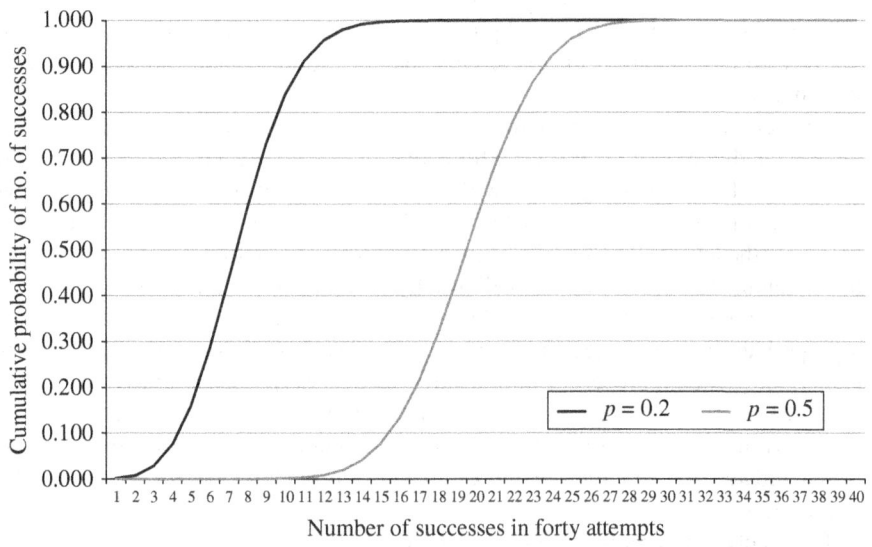

FIGURE 7.3 Binomial cumulative distributions for $p=0.2$ and $p=0.5$.

like both of these probabilities to be as small as possible. They need to choose a decision threshold for number of articles or proposals that accomplishes this.

Table 7.1 summarizes the impact of this threshold for high population probability distributions of $p=0.3$, 0.4, and 0.5. Consider a threshold of 12. PFA equals 4.3%, which means that roughly 1 of 20 low probability candidates will be promoted and tenured. If the high probability candidates can be characterized with $p=0.5$, then

MODEL OF TENURE DECISION MAKING

TABLE 7.1 Probabilities of False Acceptance (PFA) and False Rejection (PFR)

Threshold	PFA (0.2)	PFR (0.5)	PFR (0.4)	PFR (0.3)
10	0.161	0.001	0.035	0.309
11	0.088	0.003	0.071	0.441
12	0.043	0.008	0.129	0.577
13	0.019	0.019	0.211	0.703
14	0.008	0.040	0.317	0.807
15	0.003	0.077	0.440	0.885
16	0.001	0.134	0.568	0.937
17	0.000	0.215	0.689	0.968
18	0.000	0.318	0.791	0.985
19	0.000	0.437	0.870	0.994
20	0.000	0.563	0.926	0.998

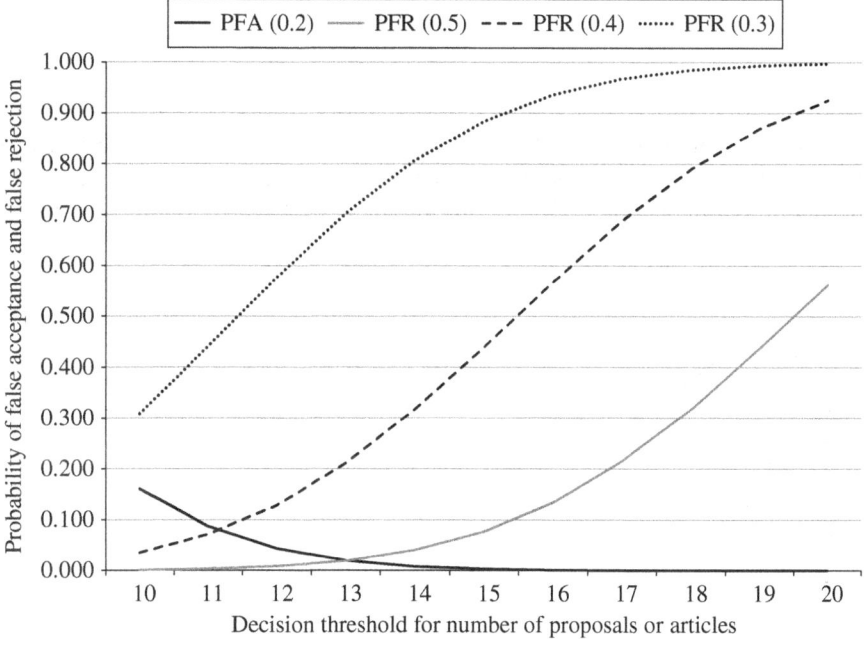

FIGURE 7.4 Probabilities of false acceptance (PFA) and false rejection (PFR).

PFR equals 0.8%, indicating that less than 1 in 100 high probability candidates will be denied promotions and tenure.

In contrast, if the high probability candidates are better characterized with $p=0.3$, then PFR equals 58%, indicating that more than one in two high probability candidates will be denied P&T. Lowering the threshold to 10 results in PFA of 16% and PFR of 31%. Clearly differentiating candidates with $p=0.2$ from candidates with $p=0.3$ is problematic. Figure 7.4 shows plots of the data in Table 7.1.

We need to transform the information in Figures 7.2, 7.3, and 7.4 into numbers tenured from low p and high p groups. To do this, we need to know how many people are in each group. To estimate these numbers, we start with the empirical probability of success from Chapter 9 denoted by p_0. For n Bernoulli trials, composed of n_1 trials with p_1 and n_2 trials with p_2, the expected value is given by

$$np_0 = n_1 p_1 + n_2 p_2 \tag{7.1}$$

We also know that

$$n = n_1 + n_2 \tag{7.2}$$

Solving for n_1 and n_2 yields

$$n_1 = \frac{n(p_0 - p_2)}{(p_1 - p_2)} \tag{7.3}$$

$$n_2 = n - n_1 \tag{7.4}$$

Using the size of each population, the number and percent tenured can be calculated for the low p population, high p population, and the overall population. The results are shown in Figure 7.5. Note that the curves for the high p population and the

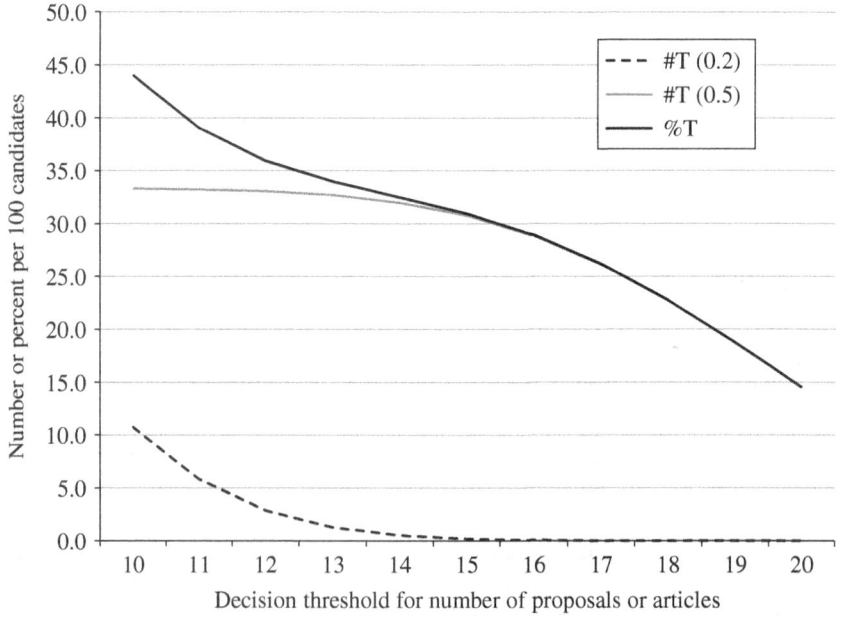

FIGURE 7.5 Number and percent tenured—low P, high P, and total.

overall population join once the decision threshold is high enough to have eliminated all of the low p population.

It is important to note that combining a $p=0.2$ population with a $p=0.5$ population to get a $p=0.3$ population means that 67% of the candidates come from the low p population and 33% of the candidates come from the high p population. Thus, the product of the these populations sizes and the probabilities from Table 7.1 yield a larger number of low p successes and a smaller number of high p successes than one might expect.

The model presented in this section enables determining percent promoted and tenured as a function of the decision threshold for differentiating low p and high p populations. In reality, there are more than just two populations. However, in my many experiences on P&T committees, I have never seen a committee successfully wrestle with more complicated formulations.

CONCLUSIONS

This chapter has addressed the nature and process of promoting and tenuring university faculty members. The nature and roles of faculty were discussed in terms of academic disciplines, faculty impact, TT versus NTT positions, availability of positions, and faculty turnover. The nature of tenure decisions was explored. I related my experiences serving on P&T committees. Finally, a model of tenure decision making was introduced that is integrated into an overall model in Chapter 12.

REFERENCES

Ashenfelter, O., & Card, D. (2002). Did the elimination of mandatory retirement affect faculty retirement? *The American Economic Review, 92* (4), 957–980.

Beecher, T., & Trowler, P.R. (2001). *Academic Tribes and Territories: Intellectual Inquiry and the Culture of Disciplines*. Buckingham, UK: Open University Press.

Breiding, A., & O'Meara, K. (2013). *Organizational Commitment and Faculty Retention Research Brief*. College Park, MD: University of Maryland, ADVANCE Program.

Carrell, S.E., & West, J.E. (2010). Does professor quality matter? Evidence from random assignment of students to professors. *Journal of Political Economy, 118* (3), 409–432.

Chronicle (2014). *The Retirement Wave: Attitudes on College Retirement and Succession Planning*. Washington, DC: The Chronicle of Higher Education.

Daly, C.J., & Dee, J.R. (2006). Greener pastures: Faculty turnover intent in urban public universities. *The Journal of Higher Education, 77* (5), 776–803.

Economist (2010). The disposable academic. Why doing a PhD is often a waste of time. *The Economist*, December 16.

Figlio, D., Schapiro, M., & Soter, K. (2013). *Are Tenure Track Professors Better Teachers?* Chicago, IL: Institute for Policy Research, Northwestern University.

Fredrickson, C. (2015). There is no excuse for how universities treat adjuncts. *The Atlantic*, September 15.

Johnsrud, L.K., & Rosser, V.J. (2002). Faculty members' morale and their intention to leave: A multilevel explanation. *The Journal of Higher Education, 73* (4), 518–542.

Kezar, A., & Maxey, D. (2014). Faculty matter: So why doesn't everybody think so? *NEA Higher Education Journal*, Fall, 29–44.

Larson, R.C., Ghaffarzadegan, N., & Xue, Y. (2014). Too many PhD graduates or too few academic job openings: The basic reproductive number R_0 in academia. *Systems Research and Behavioral Science, 31*, 745–750.

Lombardi, J.V. (2013). *How Universities Work*. Baltimore, MD: Johns Hopkins University Press.

NSF (2002). Academic research and development. In *Science and Engineering Indicators* (Chapter 5). Washington, DC: National Science Foundation.

Rosser, V.J. (2004). Faculty members' intentions to leave: A national study on their worklife and satisfaction. *Research in Higher Education, 45* (30), 285–309.

Snow, C.P. (1959). *The Two Cultures*. Cambridge, UK: Cambridge University Press.

Umbach, P.D., & Wawrzynski, M.R. (2005). Faculty do matter: The role of college faculty in student learning and engagement. *Research in Higher Education, 46* (2), 153–184.

8

EDUCATION PROGRAMS

Education is, of course, a central mission of a research university and what primarily distinguishes it from purely research institutes. Programs are usually designed to achieve specific learning outcomes, the success of which may be assessed by accrediting bodies, for example, for engineering, medicine, law, and business. Accreditation of degree programs is a requirement for students to be eligible for most federal grants and subsidies.

The accreditation process is not without critics (WSJ, 2015). It is argued that accrediting bodies act like cartels and inhibit or outright thwart innovation. Bipartisan legislation has recently been introduced that would enable the US Department of Education to influence the accreditation process. One goal of the legislation is to allow nontraditional students to use federal grants and subsidies for programs more suited to their needs.

Higher education is facing several challenges in terms of needs for talent, students' orientation toward education, and an evolving value proposition, which continues to be challenged by escalating costs. In this chapter, I first review these challenges and then address the nature of degree programs, curricula and courses, and delivery of education. I relate a variety of teaching experiences. Finally, a demand-driven faculty workforce model is presented that becomes an element of the overall integrated model discussed in Chapter 12.

Universities as Complex Enterprises: How Academia Works, Why It Works These Ways, and Where the University Enterprise Is Headed, First Edition. William B. Rouse.
© 2016 John Wiley & Sons, Inc. Published 2016 by John Wiley & Sons, Inc.

STEM CHALLENGES

In 2007, the National Academies published *Rising Above the Gathering Storm: Energizing and Employing America for a Brighter Economic Future*. The report argues that the vitality of our economy depends on "the productivity of well-trained people and the steady stream of scientific and technical innovations they produce." Globalization has increasingly made it difficult for the United States to maintain its competitive advantage in science and technology-driven innovation. This has also led to offshoring of technically sophisticated jobs, which once were solely the province of the United States.

The committee that produced this report recommended several initiatives. These included investing substantially in science and mathematics education, strengthening science and engineering research, attracting the best and brightest students from the United States and internationally, and investing in creation of high-paying jobs. They further detailed specific actions that could support achieving the goals of these initiatives.

From the perspective of this chapter, the challenge of interest is the need for a flow of science, technology, engineering, and mathematics (STEM)-capable students into the higher education system. This challenge directly relates to K-12 education, which is beyond the scope of this book. Nevertheless, it is useful to briefly review efforts to computationally model the flow of STEM students through K-12 and into higher education.

Wells et al. (2007, 2010) present an analysis and system dynamics model of student flow—both STEM and non-STEM students. They focus on student interests and proficiencies, teacher proficiencies, and class size. Kelic and Zagonel (2008) use a system dynamics model to explore boosting STEM literacy, lifting the H1-B visa cap, limiting the offshoring of jobs, and maintaining training as solutions to the declining STEM workforce. They conclude, however, that the system is very complex with many feedback loops and long time delays, so focusing on any one point of the system will not solve the problem.

The integrated model presented in Chapter 12 assumes a readily available flow of STEM-capable students from high school to higher education. For highly selective research universities, this assumption seems quite reasonable. However, for the broad academic ecosystem and the much broader economy, this assumption may be questionable. This book, however, does not address these broader issues.

STUDENT POPULATION

Beyond the STEM capabilities of incoming students, we need to understand their inclination and preferences more broadly. Van der Werf and Sabatier (2009) report on this topic based on a study sponsored by *The Chronicle of Higher Education*. Focused on the students of 2020, they reach the following conclusions:

- The traditional model of college is changing to include hybrid class schedules and much more online learning
- Four-year full-time programs will continue to wane; 4-year graduation rates have continued to decline

- Students' convenience will be a priority so they can balance education, work, and family commitments
- Online programs will have to expand quickly so that both on-campus and off-campus students can be accommodated
- All of the aforementioned options will have to be added while continuing traditional education for those wanting the overall on-campus experience
- To address the increasing numbers of ill-prepared students, colleges may add a remedial year prior to enrollment in bachelors programs
- Around 2020, minority students on campus will outnumber whites, steadily shifting needs, preferences, and orientation
- The pressure on colleges to adapt will increase as traditional and nontraditional competitors adapt to meet these needs

In my experience, many current students exhibit these characteristics. For several years, I taught one class of a predictive health course at Emory University. There were usually about 100 undergraduates in this elective class. The lead instructor finally had to prohibit laptops in the class to keep students from multitasking during class. For example, while I was lecturing and doing my best to make it interesting and perhaps entertaining, students would be shopping on retail websites and keeping up with Facebook posts.

For the last 3 years, I have each semester been one of the instructors for an undergraduate engineering economics course required of every student on campus. I lecture on five advanced topics in every section of the course, thereby interacting with every junior on campus. I have found that I get greater student engagement when I tell stories of my use of engineering economics to address real problems in industry. I often get applause after these lectures.

We recently tried another approach to enhancing engagement. At the beginning of the lecture, I announced that there would be a brief quiz at the end of the lecture, limited to the content of the lecture they had just heard. For the first time in six semesters of teaching this course, I noticed that not a single student was on their laptop, tablet, or smartphone. They were engaged because their learning would be evaluated in 30–40 minutes.

Unlike previous generations, the current generation of students has a wealth of very engaging ways to allocate their attention. Teachers have to compete with these alternatives, either directly if students have access to their devices during class or indirectly due to students' overall increased expectations of engagement. This makes the performance aspects of teaching more important, which includes paying careful attention to what is working, at the moment, and what is not. I return to this topic later in this chapter during discussions of the delivery of education.

VALUE OF EDUCATION

In several earlier discussions, I addressed the impact of escalating costs of higher education and, in some areas, decreasing job prospects for college graduates. At least in the popular press, there are increasing questions of the economic value of education.

Shribman (2011) asks the question, "Does formal education matter?" He conducted an informal survey of several college presidents asking whether Abraham Lincoln would have been a better president if he had a college education. There was complete agreement among those asked that Lincoln became educated through his own efforts rather than formal education. Thus, no institution certified his education. I discuss in the following the value of a degree as certification of knowledge and skills.

Of course, Shribman's question is a bit of a red herring. We do not have the luxury of most people being comparable in talent and motivation to Abraham Lincoln. His point, however, is that higher education may not really be essential. Harris (2011) takes this a step further. He questions whether higher education is worth the investment.

Harris starts by noting that in August 2011, student loans surpassed credit cards as the nation's single largest source of debt, edging ever closer to $1 trillion. Yet, he argues, "For all the moralizing about American consumer debt by both (political) parties, no one dares call higher education a bad investment. The nearly axiomatic good of a university degree in American society has allowed a higher education bubble to expand to the point of bursting."

He further notes that "Wages for college-educated workers outside of the inflated finance industry have stagnated or diminished. Unemployment has hit recent graduates especially hard, nearly doubling in the post-2007 recession. The result is that the most indebted generation in history is without the dependable jobs it needs to escape debt."

He concludes, "If tuition has increased astronomically while the portion of money spent on instruction and student services has fallen, and if the market value of a degree has dipped and most students can no longer afford to enjoy college as a period of intellectual adventure, then at least one more thing is clear—higher education, for-profit or not, has increasingly become a scam." Harris, obviously, feels pretty strongly about this.

The key point relative to the evolving value proposition of higher education is that there are increasingly strident commentaries regarding declining value, emerging bubbles, and so on. Clearly, universities would be remiss, not to mention rather arrogant, if they ignored these critiques. Chapters 11 and 12 address such needs to change in great detail.

DEGREE PROGRAMS

In principle, everyone could, like Abraham Lincoln mentioned earlier, educate themselves. It would be amazingly inefficient, but more importantly there would be no way for high schools, undergraduate schools, and graduate schools to know what incoming students knew. Employers would have the same problem assessing graduates. The credential of a degree, and often a transcript, helps to solve these problems.

Undergraduate programs (B.S. or B.A.) are usually the largest programs at most universities, although a few research universities have more graduate students than undergraduates. The bachelor's degree was intended to be completed in 4 years but increasingly takes 5 or 6 years. I have heard repeatedly that universities try to break even economically on undergraduate programs.

The master's programs (M.S., M.A., M.F.A.) tend to be profitable, particularly if they are provided to corporate customers who pay the tuitions of students. It is quite common for master's level courses to be rather large and taught by nontenure-track faculty members. Thus, revenue is high due to large classes and costs are low due to use of teaching, adjunct, or part-time faculty members. Such programs are often the "cash cows" of a department or school's offerings.

Doctoral programs (Ph.D., D.Sc.) are labor intensive with smaller classes, tenure-track faculty, and a research advising process that limits the number of doctoral students per faculty member. Typical faculty members may have three to four active Ph.D. students. I have had as many as 10–12 but all of them were coadvised by junior faculty members. This larger number can work by spacing the students out and having senior Ph.D. students mentor junior Ph.D. students.

Many universities allow part-time Ph.D. students who are usually working full-time while pursuing their studies. It is not surprising that progress is often erratic and such students, if they complete their degree, typically take a much longer time than full-time students. Nevertheless, many of these students are very competent and highly motivated and do well, particularly when their corporate employers are supportive, both organizationally and financially.

The portion of science and technology full-time Ph.D. students from the United States has steadily decreased. These people are readily employable with attractive salaries. The idea of giving up $6000–8000 (or more) per month for a Ph.D. stipend of $2000 per month is very difficult for US citizens to embrace. Consequently, foreign students, who readily accept the $2000 monthly stipend plus tuition as compensation, dominate full-time Ph.D. ranks in science and technology.

CURRICULA AND COURSES

The faculty is responsible for the design of the curricula and courses, as well as delivery of courses. The curriculum for a particular degree within a department is usually developed, and regularly updated, by the undergraduate and graduate curriculum committees. All faculty members of the department vote on the proposed new or modified curriculum. If approved, the curriculum is recommended to the college or school curriculum committee. If approved at that level, it goes to a university-level committee.

One or more faculty members usually design courses. These faculty members will have expertise in the topic of the course. Courses proposals are reviewed and voted upon by the appropriate curriculum committee. They then go to the college and school level and perhaps the university level. Finally, the course information goes to the function that maintains the course catalog and schedules the delivery of courses.

Faculty members often "own" one or more courses in a program and frequently teach these courses, which obviously highly influences course content and delivery. For new faculty members, preparation of course notes often dominates their time. This can be quite demanding for younger faculty members who are also trying to build a research program. When both members of a couple are young faculty members and they have young children, the early years in their careers can be quite stressful.

At Illinois, I taught undergraduate and graduate courses in control theory and graduate courses in decision theory and modeling human–machine systems. I continued teaching these two graduate courses at Georgia Tech. I also taught a few unusual courses as described in the next section. At Stevens I teach undergraduate engineering economics and graduate modeling and visualization of complex systems and enterprises.

Developing and delivering courses is a great way to learn the material in depth. When I was at Illinois in the late 1970s, I volunteered to teach graduate courses on Design of Experiments and Queuing Networks. I had never had courses on these topics, nor their prerequisites, as a student and I wanted to learn these topics in depth, as they were central to my research. I had to work very hard to stay ahead of the graduate students. Fortunately, I taught these courses in different semesters.

O'Connell (2015) reports on a very creative course offering. Stanford's most popular class is not in computing. It's called "Designing Your Life," a course that's part throwback, and part a foreshadowing of higher education's future. "Designing Your Life" is a new and wildly popular course for Stanford juniors and seniors that is grounded in design thinking concepts and techniques. The course's lessons give students the perspective they need to navigate decisions about life and work postgraduation. They learn gratitude, generosity, self-awareness, and adaptability.

DELIVERY OF EDUCATION

Robinson (2010) argues, quite compellingly, for changing education paradigms. He characterizes traditional education as production-oriented delivery where every student of the same age is taught the same material. They are taught by boring instructors who stand up front and talk at them while tediously writing lecture material on the board as they have been doing for many years. Computer-generated presentations are better but do not address the fundamental problem.

The current generation of students has portable computers, tablets, and smartphones. These students are always connected, know how to access information quickly, and interact with each other constantly. They multitask incessantly, pursuing schoolwork, perhaps a job, and social life all at once. Consequently, production-oriented education no longer works.

Much of university education can be characterized as "Sage on the Stage," but this is slowly changing to "Guide on the Side." There is currently much interest in "flipping the classroom," whereby students are, in part, responsible for course delivery. I have experimented with this three times in recent years.

The first time was during the summer, when I seldom teach. Several graduate students approached me about organizing a summer course on Smart Grid, a phrase used to describe the next generation electric grid. I told them that I would organize the material but they would have to teach it. Each student was responsible for one 2½ hour period, which was roughly half presentation and half discussion.

We ended up with about 10 presentations and sets of discussion notes. The last assignment was to update all the presentations, based on discussion comments and suggestions, and then integrate them all into a single 1-hour presentation. The goal was a single coherent and compelling intellectual product that each of us, including me, could use in the future. The students were quite pleased with the outcome.

The next opportunity came when several students who had taken my graduate decision theory course asked me to teach a follow-up course. I decided to make the course case based and used Kershaw's (2007) *Fateful Choices: Ten Decisions that Changed the World*. The decisions were all associated with World War II and were made by Winston Churchill, Adolf Hitler, Benito Mussolini, Franklin Roosevelt, Joseph Stalin, and Hideki Tojo. Each chapter of this fascinating book described decision makers' objectives, what alternatives they had, what information was available, and the associated uncertainties and risks.

Each student was responsible for one decision. They had to use the models and methods from my earlier course to formulate each leader's decision using multiattribute utility models. Their presentations included a discussion of their case, their formulation of the decision, and their assessment of the extent to which the decision was rational, albeit not necessarily admirable. The students did great jobs. One student said it was the best history course he had ever taken.

The third experience involved designing a student-taught course called "Transforming Academia." The experience of leading this course strongly influenced my writing of this book. The course included eight Ph.D. students and four faculty members. Each student taught two 1–2-hour modules. The content covered the 1000 years of higher education discussed in Chapter 2.

As with the Smart Grid course, I required the students to integrate their materials into a single 45-minute presentation. This integrated intellectual product was presented to Georgia Tech's President and Provost, with emphasis on their recommendations for transforming Georgia Tech. The presentation and discussion lasted 2 hours, with many good insights and ideas emerging. Several students commented later that this course was a very valuable experience.

Beyond flipping the classroom, technology-enabled course delivery has become increasingly prevalent, ranging from immersive technology environments in the classroom to remote web-based delivery. DeMillo (2011, 2015) provides insightful discussions of the possibilities. The future scenarios discussed in Chapters 11 and 12 will challenge many of the common course delivery practices and elaborate the need to leverage such technologies.

TEACHING EXPERIENCES

In my first year of teaching, at Tufts University in 1973, a foreign graduate student came to my office to complain. I had provided a problem statement in an operations research exam that included much more information than necessary. Some of this extra information was irrelevant. My goal was to get students to formulate problems amidst too much information.

The student complained that I had not done what his professors did in his native country. When they gave word problems, they underlined the information you needed to solve the problem. I sympathized with him but did not agree. I told that real-life problems do not come with the key information underlined.

In my graduate course on decision theory and decision support systems, I have always liked making at least one question on the midterm or final exams open ended. Examples include:

- Despite having the requisite technology, Motorola delayed introducing digital cell phones to avoid cannibalizing their analog advantage. Nokia and Qualcomm introduced digital cell phones and gained tremendous competitive advantage. Under what decision theory assumptions was Motorola's decision rational?
- Determine the optimal level of the federal stimulus funds during the Great Recession and the optimal allocation of these funds across the financial, production, and consumption sectors of the economy
- Consider the university budget, taking into account all the key stakeholders in the various functions of the university enterprise. Determine the optimal allocation of the university budget across education, research, and service

These were all take-home exams for a 2-week period, with in-class discussion of difficulties encountered after the first week.

I have often been amazed at the quality of answers provided by a few students. At the same time, I have found many students have a tough time getting started. That is why I added the in-class discussion after the first week. If the problem assigned is designated a Type X problem, most students do pretty well. On the other hand, if it is not at all clear what type of problem it is, they may struggle.

This is central in senior design, the typical capstone project course where students address a real-world problem for real sponsors who want the results they will provide. Examples of projects are:

- The average time in the emergency room has increased from 4 hours to 5. What has caused the 1-hour increase and how can this be remediated?
- The average time from the doctor signing the discharge order until the patient actually leaves the bed is 7 hours. How can this time be decreased?
- The utilization of the 28 operating rooms is at capacity. Can this utilization be better managed to, in effect, create more capacity or are more operating rooms needed?

Advising teams of four to six students addressing problems such as these has always been quite interesting. Beyond discussing the context of the problem, our weekly meetings focused on formulating messy problems, finding data to clarify ambiguities, and making sure each team member was playing an important role and contributing. To this end, three times during the semester, team members rated each other's contributions.

At Illinois, I taught a required senior course on control theory six semesters in a row. This topic is fairly mathematical with each concept and formulation building on previous ones. On the last day of this class, a senior who had sat in the front row all semester says, "I don't get it." I asked what aspect of this last lecture he did not get. He said, "I have not understood anything the whole semester." We had a bit of class discussion on the nature of the course material and what aspects many found difficult. He passed the final exam, got a C in the course, and accepted an industry job in control systems!

I have been teaching undergraduate engineering economics for the past six semesters, having not taught undergraduates for 35 years. I spend half of the first lecture outlining the set of five lectures in terms of the concepts students will learn as well as who they will meet, for example, five winners of the Nobel Prize in Economics. Before each lecture, I present one slide that explains the problems they will be able to address using the material I present in this lecture and the other lectures. In other words, I answer the question, "Why are you telling us about this?"

Sometimes when your attention is captured by some tough research question, you can lose track of the primary function of the university. I have had many industry research sponsors at Illinois, Georgia Tech, and Stevens and all of them tell me the same thing. When I ask them if they feel our research outcomes are of benefit to them, they say something like, "Yes, your research is important to us, but what is most important is your students who we want to hire when they graduate."

At the graduate level, many students are already employed by industry and the companies involved want our education program to enhance their knowledge and skills of relevance to the company. I have been involved in several efforts to adapt delivery methods to better accommodate such students. Beyond distance learning, courses have been designed to be delivered all day Friday and Saturday once per month, or all in one 40-hour week. These changes have proven to be quite successful.

One particular change has been very difficult. The idea of a "course" is deeply ingrained in academia. This includes the notion that you enroll in the course during a particular semester and complete it during that semester. This is sacred but often does not make sense. As discussed earlier, we developed a professional master's degree composed of roughly 100 four-hour modules. We were forced by the Registrar to organize these modules into courses that were completed in one semester because they could not figure out how to charge for modules. They told us that their IT system would not allow it.

Many universities pride themselves on their selectivity. At an extreme, a director of a program at a university where I was on the advisory board touted the fact that his program was so selective that they only admitted one Ph.D. student per year. I suggested that he could be even more selective by admitting no one. He said that there would then be no program. I suggested that he did not have a program now, just one very good student.

This leads me to an observation about Ph.D. education. Somewhat simplistically, I have encountered two educational philosophies at the five universities where I been a faculty member. One philosophy asks the Ph.D. student to continually prove why they should be in the program. Course exams and particularly qualifying and comprehensive exams are intended to weed out people who cannot prove they are worthy. The role of faculty members is to assess whether students are good enough to continue.

The other philosophy, which I have adopted, is to be very careful in the selection process to understand students' backgrounds, aspirations, and work ethics. Once they are admitted and choose to enroll, my role is to help them succeed. They still have to jump all the same hurdles but my job is to serve as a mentor and coach to help them understand, prepare for, and succeed at these feats. Some will still fail, but not because I calmly watched it happen.

This philosophy presents some difficulty for foreign students because I will not agree to work with a student until I get to know them. Thus, they may have to enroll without assurances of financial support. However, I am not just looking for perfect GRE scores and a transcript with only top grades. If I am to invest roughly 4 years of at least 1 hour per week, and at key points much more, working with a Ph.D. student, he or she and I need to have aligned aspirations and compatible work ethics.

WORKFORCE MODEL

Educational programs drive recruitment of faculty members to teach the courses in these programs. Faculty members can be both tenure track and nontenure track. Tenure-track members have lower teaching loads, typically around 50% of the teaching load of nontenure-track faculty members. The reduced teaching load is to enable tenure-track faculty members to pursue the research that will earn them tenure and promotions. This requires that write proposals to secure funding and publish journal articles, book chapters, and books that lead to citations and renown. Chapter 9 addresses these research activities.

In this section, I will consider how to project the number of faculty—the workforce—needed by the university enterprise over time. The workforce model requires several inputs to project both undergraduate and graduate student populations, numbers of classes, and number of faculty needed:

- Initial size of each student population
- Growth rate of each student population
- Classes per semester for each type of students
- Students per class for each type of students
- Teaching loads for tenure-track faculty members (TTL)
- Teaching loads for nontenure-track faculty members (NTTL)
- Percent tenure-track faculty members (%TT)

With these inputs, it is straightforward to calculate number of classes (NC) offered per semester over time. The total number of faculty (NF) needed is then given by

$$NF = \frac{NC}{\left[\%TT \times TTL + (1-\%TT) \times NTTL\right]} \tag{8.1}$$

The number of tenure-track faculty members (#TT) and number of nontenure-track faculty members (#NTT) are given by

$$\#TT = \%TT \times NF \tag{8.2}$$

$$\#NTT = (1-\%TT) \times NF \tag{8.3}$$

These calculations provide the #TT and # NTT for each of the years being projected. We now need to distribute these numbers over junior and senior faculty members. These can be accomplished by defining an array of faculty members where the ages of faculty members are the rows and the years into the future are the columns. Thus, a faculty member of age T at year t becomes a faculty member of age $T+1$ at year $t+1$. We need to do this to represent faculty retirements.

Beyond having faculty members get older, we also need to include three other phenomena:

- After 6 years, faculty members are either granted tenure or not. In the workforce model, the tenure model from Chapter 7 is used
- Once tenured, faculty turnover is represented by a fixed percentage for each age. The example presented in the following assumes 5%, but any percentage can be chosen
- Once 65 years old, faculty members retire with a fixed percentage each year until age 80. The example in the following assumes 10%, but any percentage can be chosen

The population of new tenure-track assistant professors is calculated to provide the targeted #TT each year. From my experience the targeted #NTT is much easier to provide due to a much larger pool of acceptable candidates. Thus, the process of hiring such faculty members is not included.

To illustrate the use of the workforce model, the scenario includes the following inputs:

- 2500 undergraduate students with an annual growth rate of either 2 or 4%
- 3500 graduate students with an annual growth rate of either 4 or 8%
- Four classes per semester for undergraduates and two for graduate students
- 20 students per class for undergraduates and 10 for graduate students
- Two courses per semester for tenure-track faculty members
- Four courses per semester for nontenure-track faculty members
- 50% tenure-track faculty members

The low growth scenario begins with 183 tenure-track faculty members in year 1 and grows to 332 faculty members in year 20. The high growth scenario has the same beginning but grows to 607 tenure-track faculty members in year 20. Figures 8.1 and 8.2 show the impact of differing tenure percentages (p) of junior faculty members for the low growth scenario. Figures 8.3 and 8.4 show the impact of p for the high growth scenario.

These projections assume that all new faculty members are hired as assistant professors. The lower the tenure percentage, the more of these faculty members will have to be replaced 6 years later. For $p=0.3$, the percent of nontenured tenure-track faculty reaches 67%.

Not surprisingly, the situation gets worse for the high growth scenario. The percent of nontenured tenure-track faculty reaches 74% for $p=0.3$. In contrast for $p=0.7$, this percentage is 45% for the low growth scenario and 55% for the high growth scenario.

The implications of this example are clear. A high growth rate coupled with a low tenure rate results in far too many nontenured tenure-track faculty members. The likely consequence is that all of the tenured tenure-track faculty would have to populate the many committees and perform other service activities associated with rapid growth. Morale would likely suffer and the turnover rate would increase from the assumed 5% noted earlier, which would serve to exacerbate the overall situation.

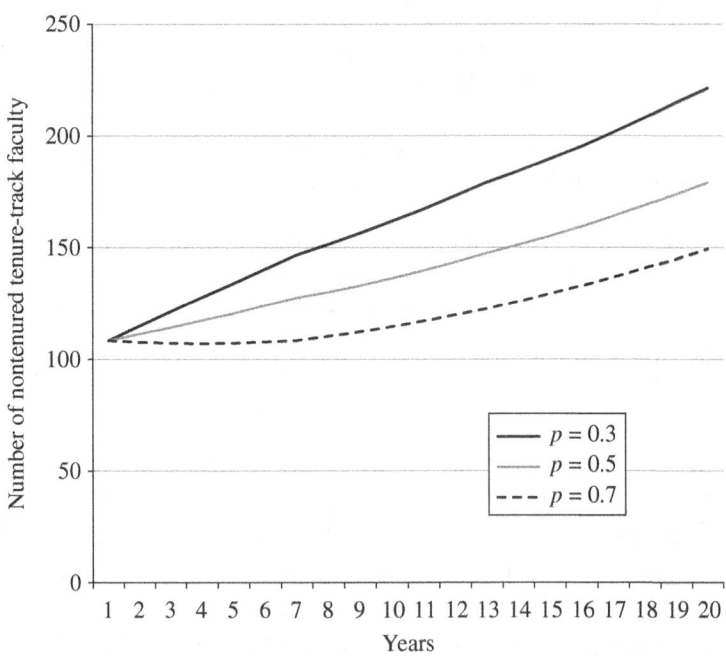

FIGURE 8.1 Number of nontenured tenure-track faculty members (low growth scenario).

WORKFORCE MODEL

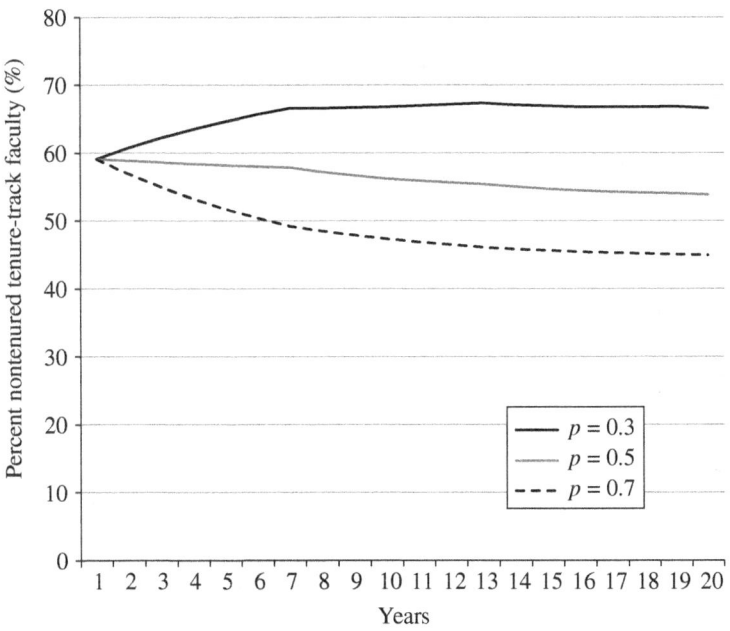

FIGURE 8.2 Percent nontenured tenure-track faculty members (low growth scenario).

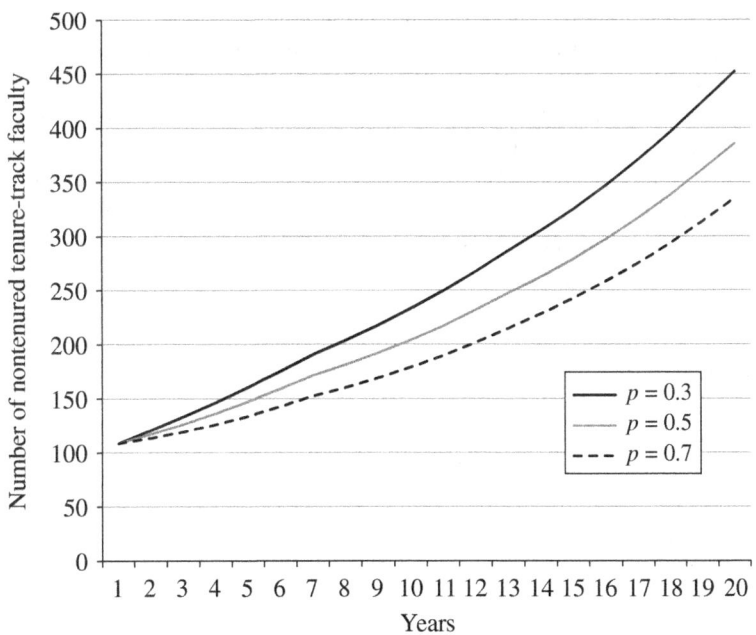

FIGURE 8.3 Number of nontenured tenure-track faculty members (high growth scenario).

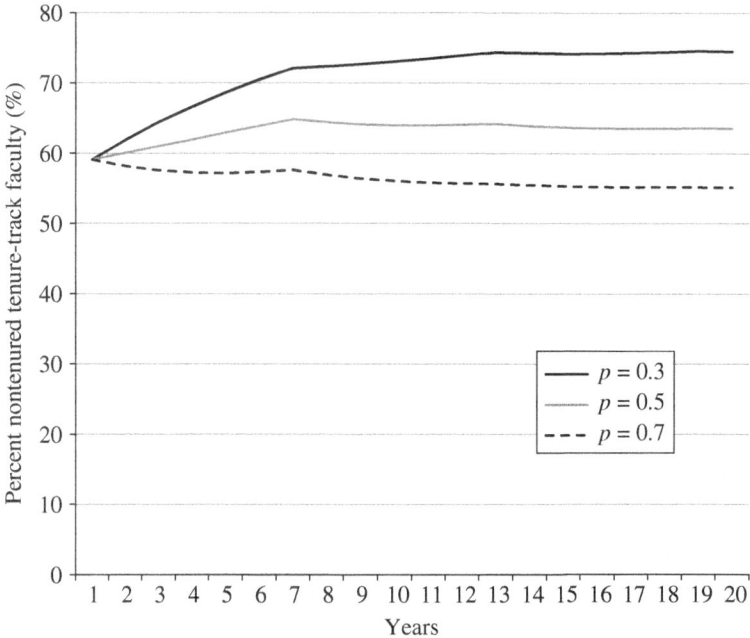

FIGURE 8.4 Percent nontenured tenure-track faculty members (high growth scenario).

A possible solution is to recruit more senior faculty members. They are more expensive and more difficult to recruit. They also demand higher salaries and monies for laboratories and equipment. The peak recruiting year for the high growth scenario and $p = 0.3$ targets hiring 90 new assistant professors. Perhaps 10–15 of these could be replaced by senior hires, but probably not many more.

Another approach to resolving this problem is to decrease %TT. However, as discussed in Chapter 10, this strategy would undermine brand value because fewer faculty members would be publishing, getting cited, and becoming famous. Thus, yet another trade-off emerges that is addressed in later chapters.

CONCLUSIONS

This chapter has addressed the current and future student populations that higher education will serve. Both STEM challenges and evolving student inclinations and preferences were discussed. Challenges to higher education's value proposition were reviewed. Types of degree programs and the design and delivery of curricula and courses were considered, including alternative approaches to delivery. I related a variety of my teaching experiences. Finally, a workforce model was introduced and a range of staffing trade-offs were discussed.

REFERENCES

DeMillo, R.A. (2011). *Abelard to Apple: The Fate of American Colleges and Universities.* Cambridge, MA: MIT Press.

DeMillo, R.A. (2015). *Revolution in Higher Education: How a Small Band of Innovators Will Make College Accessible and Affordable.* Cambridge, MA: MIT Press.

Harris, M. (2011). Bad Education. *N+1 Magazine,* April 25.

Kelic, A., & Zagonel, A. (2008). *Science, Technology, Engineering, and Mathematics (STEM) Career Attractiveness Systems Dynamics Model.* Albuquerque, NM: Sandia National Laboratories.

Kershaw, I. (2007). *Fateful Choices: Ten Decisions That Changed the World.* New York: Penguin.

National Academies (2007). *Rising Above the Gathering Storm: Energizing and Employing America for a Brighter Economic Future.* Washington, DC: National Academy Press.

O'Connell, A. (2015). My creative life. *Fast Company,* April 24.

Robinson, K. (2010). Changing Education Paradigms. http://www.youtube.com/watch?v=zDZFcDGpL4U (accessed February 8, 2016).

Shribman, D. (2011). Does formal education matter? *Real Clear Politics,* March 2011.

Van der Werf, M., & Sabatier, G. (2009). *The College of 2020: Students.* Washington, DC: Chronicle Research Services.

Wells, B.H., Sanchez, A., & Attridge, J.M. (2007). *Systems Engineering the U.S. Education System.* Waltham, MA: Raytheon.

Wells, B.H., Sanchez, A., & Attridge, J.M. (2010). *Modeling Student Interest in Science, Technology, Engineering, and Mathematics.* Waltham, MA: Raytheon.

WSJ (2015). Trust busting higher ed. *Wall Street Journal,* October 5, A18.

9

RESEARCH AND INTELLECTUAL PROPERTY

The research reputations of faculty members, and universities as a whole, affect how others rank your programs and the quality of the students who apply for admission. For the 4 years when I was a voter for *US News & World Report*, my rankings were based on the faculty members that I knew at each university and my opinion of their research. To a great extent, it would have been very difficult to make this assessment any other way.

This chapter first addresses the challenges of research in terms of the role of peer review of publications and proposals, bibliometrics associated with citations, and difficulties securing funding. I then discuss a variety of research experiences. The subsequent section presents a model of research that predicts probabilities of articles being accepted, expected citations after publication, and probabilities of research proposals being funded. This model is a central element of the integrated model discussed in Chapter 12.

The last major section of this chapter addresses intellectual property (IP), which is often one of the outcomes of research. In some cases, this takes the form of patents, but more often it is embodied in the know-how of the researchers. Either form of IP may result in spin-off businesses that create jobs, revenue, profit, and perhaps public stock offerings. I relate several of my experiences in launching and growing spin-offs.

Universities as Complex Enterprises: How Academia Works, Why It Works These Ways, and Where the University Enterprise Is Headed, First Edition. William B. Rouse.
© 2016 John Wiley & Sons, Inc. Published 2016 by John Wiley & Sons, Inc.

CHALLENGES

The essential challenge for tenure-track faculty members is to create a track record that will earn them promotion and tenure as discussed in Chapter 7. Two major portions of this track record are articles accepted for publication and proposals funded. Central to accomplishing this is peer review.

Peer Review

Articles submitted to journals and proposals submitted to potential sponsors are reviewed, usually anonymously, by disciplinary peers who may, or may not, recommend publication or funding. Peer review is seen as a core value of the academic enterprise.

Uzoka et al. (2013) present a multicriteria framework for assessing scholarship. They surveyed faculty members in five Canadian universities to assess the weights that they apply to different elements of faculty members' education, research, and service activities and accomplishments. Not surprisingly, the faculty members surveyed valued scholarship of discovery that resulted in publications in peer-reviewed journals with high impact. Other activities were much less valued.

Harley and Acord (2011) address the meaning, locus, and future of peer review. They stress "the need for a more nuanced academic reward system that is less dependent on citation metrics, the slavish adherence to marquee journals and university presses, and the growing tendency of institutions to outsource assessment of scholarship."

They also note that the tenure and promotion requirements of elite universities have trickled down to less competitive and nonresearch-intensive institutions. "The global effect is a growing glut of low quality publications that strains the efficient and effective practice of peer review, a practice that is, itself, primarily subsidized by universities in the form of faculty salaries."

Harley and Acord also consider the impact of the "open-access" trend on peer review. It will take quite some time before these new journals with new formats and review processes will displace the aforementioned marquee journals. For some programs, such as business schools, the almost religious adherence to publishing in marquee journals will be difficult to displace. As discussed in Chapter 11, however, the future may be quite different than the present.

Bibliometrics

If all goes well, articles are accepted and published and are then cited by peers. One's citation record is a primary means for judging the impact of one's publications. The study of patterns of citations is an element of the field of bibliometrics. Pritchard (1969) defined bibliometrics as "the application of mathematics and statistical methods to books and other media of communication."

In a classic NSF-funded report, Narin (1976) discusses how libraries first used citation counts in the early 1900s to evaluate adequacy of their collections. Evaluative bibliometrics uses publications and citations as indicators of individual and organizational productivity and eminence. While this practice is not without controversy—see following text—Narin notes that "Most bibliometric evaluations of papers, people, or institutions correlate well with peer evaluations."

Saracevic (2004) reviews several "laws" and key developments within bibliometrics including:

- Lotka's law: The total number of authors y in a given subject, each producing x publications, is inversely proportional to some exponential function n of x. $n = 2$ for science, so $x^2 y =$ constant
- Bradford's law: n journals produce N articles, the next n^2 journals produce N articles, the next n^3 journals produce N articles, and so on
- Zipf's law: The frequency distribution of words in articles follows the relationship rank x frequency = constant
- Gosnell obsolescence: Concerns the half-life of citations of a journal article, which vary by discipline
- Garfield's Science Citation Index in 1961 and origination of the Impact Factor

These laws emerged from many decades of study of patterns of both publications and citations.

Bornmann and Mutz (2015) report on growth rates of modern science via a bibliometric analysis based on the number of publications and cited references. They report that the annual growth rate of publications and citations was 1% from the mid-1600s to the mid-1700s, 2–3% between the two World Wars, and 8–9% in 2012. Ware and Mabe (2012) report that there were 28,100 peer-reviewed journals in 2012, collectively publishing 1.8–1.9 million articles per year. The number of articles published grows by 3% per year, number of journals by 3.5%, and number of researchers by 3%.

They also note that "Journals do not just disseminate information; they also provide a mechanism for the registration of the author's precedence; maintain quality through peer review and provide a fixed archival version for future reference." Thus, the tsunami of journal publications plays an important role in the research enterprise.

Ioannidis (2014) addressed the question, "Is your most cited work your best?" He surveyed 400 top-cited authors, achieving a roughly 33% response rate. Most top-cited authors agreed that their best papers were the highly cited ones. They also indicated that top-cited papers tend to be more evolutionary than revolutionary. In other words, the most highly cited papers fit into the reigning paradigms.

Werner (2015) argues that the focus on bibliometrics makes papers less useful. He argues that "Many of the negative effects of bibliometrics come not from using it, but from the anticipation that it will be used." For example, researchers try to package their work within the limits imposed by *Nature* and *Science* rather than using less prestigious outlets that will accommodate, for instance, much longer expositions.

Finally, it is useful to note Thomson-Reuters (2008) *Using Bibliometrics: A Guide to Evaluating Research Performance with Citation Data*. This guide provides an overview of Web of Science, an online tool, and citation data sets, for valuation of research performance. Keep in mind, however, the valuation problem explicated in

Chapter 7 where inherent randomness can make it difficult to discriminate high performers from low performers without accepting significant probabilities of false acceptances and false rejections.

Funding

Faculty members need time and the help of Ph.D. students and postdocs to accomplish the research that enables preparing and submitting articles. Faculty members' summer salaries, as well as stipends for Ph.D. students and postdocs, need to be funded. Such funds are typically secured by submitting proposals to research agencies, foundations, and industry.

Editors (2011) explains how this process is getting more and more difficult. "The process of applying to government agencies and private foundations for grants has become a major time sink. In 2007 a U.S. government study found that university faculty members spend about 40 percent of their research time navigating the bureaucratic labyrinth, and the situation is no better in Europe. An experimental physicist at Columbia University says he once calculated that some grants he was seeking had a net negative value: they would not even pay for the time that applicants and peer reviewers spent on them."

They note that "A vicious cycle has developed. With more and more people applying for each grant, an individual's chances of winning decrease, so scientists must submit ever more proposals to stay even. Between 1997 and 2006 the National Science Foundation found that the average applicant had to submit 30 percent more proposals to garner the same number of awards. Younger scientists are especially hard-pressed: the success rate for first-time National Science Foundation applications fell from 22 percent in 2000 to 15 percent in 2006."

They also describe some alternative funding models. Howard Hughes Medical Institute, the largest private supporter of medical research in the United States, is providing larger grants for longer periods with much less red tape. The Wellcome Trust in the United Kingdom is now shifting to a similar system.

Howard and Laird (2013) discuss the new normal in funding university science. They plow some of the same ground, but a bit more deeply. Success rates at NIH and NSF have fallen from 30% in 2001 to 20% or even less in 2011; during roughly the same period, the number of proposals submitted doubled. Growth of federal funding has not kept up with the growth of academic research enterprises. Sixty years of such growth has led to an unsustainable enterprise at a tipping point.

The deeper source of the problem, they argue, is the incentive structure of research universities. Increasing emphasis on research requires expensive personnel to assist faculty members and, hence, doctoral programs and increasingly postdoctoral programs. The trend can be seen in Ph.D. graduation data:

- 1920–1924: 545 science and engineering doctorates per year
- 1955–1959: 5662
- 1995–1999: 26,854
- 2010: 27,134

At the same time, nondefense R&D funding by the federal government has been around 10% of the discretionary portion of the federal budget for 40 years. Institutional funding of research has risen from 12% in 1972 to 20% in 1991, remaining at that level through 2009. Industry funding has fallen from 7–8% in the 1950s to 5% in 2011. Increasing such funding will require deeper understanding of the needs of both business and academia, as discussed in the next section.

The bottom line of this section is that it is getting more and more difficult to publish in top journals and secure funds from highly regarded sponsors. The annual growth rate of demand is enormous, bordering on 10% for journal submissions and 5% for proposals submissions. As will be discussed in the context of the research model later in this chapter, this yields a situation where faculty members have to work harder and harder to achieve less and less success.

RESEARCH EXPERIENCES

In this section, I relate various research experiences that illustrate the serendipitous nature of how research ideas emerge and are developed, as well as the preferences and inclinations of various research sponsors.

Libraries and Networks

During 1972–1979, while at MIT, Tufts, and Illinois, I conducted a variety of studies on the use of operations research methods to improve operations of libraries, as well as networks of libraries. This research was prompted by fascination with Philip Morse's *Library Effectiveness* (1968) that showed how relatively simple models can be used to provide valuable insights into how to improve operations, in this case of libraries.

Morse's book has served as a model for me relative to how to approach problems and keep models both tractable and understandable by managers of complex organizations. Our recent book on healthcare delivery (Rouse & Serban, 2014) was written with Morse's model in mind. Indeed, this book on universities as complex enterprises follows Morse's model as well.

Limits of Modeling

I spent 1979–1980 at Delft University of Technology. In an effort to improve my Dutch, which I was required to learn, I mentioned earlier that I sat in on faculty and staff discussions of fundamental limits of sports. This led me to think about fundamental limits in physics, chemistry, biology, etc. This, in turn, led me to conduct a series of studies of fundamental limits in modeling human cognition.

At that time, we had undertaken a major project to develop an artificially intelligent copilot for military fighter pilots—this is detailed later in the discussion of spin-offs. This AI copilot had to "understand" the pilot. This was accomplished by representing pilots' goals, plans, and scripts, as well as information and control requirements, associated with the various activities of piloting. We applied the notion of fundamental limits to evaluate the validity of the knowledge embedded in this representation. It turned out to be a useful way to uncover knowledge engineering errors.

Healthcare Delivery

I spent 1973 as a visiting assistant professor at Tufts. I was involved in a research project to automatically control radiotherapy equipment at the Tufts New England Medical Center. A freshly minted MIT Ph.D., the MDs at the Center treated me like a well-educated technician. I avoided being involved with healthcare research for 25 years after that experience.

In the late 1990s, I undertook a variety of efforts for the executive team at the American Cancer Society. This was an entirely different experience. Senior medical colleagues treated me like a colleague who had skills complementary to theirs. I immersed myself in the complex nature of disease control.

During the mid- to late 2000s, I became heavily involved in the National Academies' healthcare delivery initiatives. Thought leaders from the Institute of Medicine, now called the National Academy of Medicine, asked thought leaders from the National Academy of Engineering to engage in engineering a learning healthcare system. Several efforts ensued over roughly a decade.

My involvement in these efforts enabled me to meet several remarkable people who mentored me on the nature of the healthcare delivery ecosystem. This led to an ongoing series of research projects with academic health centers at Emory University, Vanderbilt University, Indiana University, and the University of Pennsylvania. The four decades since my experiences at Tufts have very much changed my view of healthcare delivery research.

Interactive Visualization

We developed in 2010 a computational model of Emory University's prevention and wellness program for their employees. Our primary motivation was to demonstrate to National Academy members that such modeling was feasible and useful. It quickly became apparent, however, that many in this cohort of very smart people did not really understand the model.

This realization led us to develop several interactive visualizations so that anyone present could take the controls and fly what came to be known as a policy flight simulator. They were able to vary assumptions and redesign the organizational model whereby the elements of this prevention and wellness program were delivered. It was rather amazing how compelling most people found this to be.

Soon after I moved to Stevens in 2012, we developed plans for the *Immersion Lab*, an 8' by 20', 180°, and touch-sensitive display system, with supporting computational infrastructure. The lab was developed to provide senior decision makers with experiences of being immersed in the complexity of enterprises ranging from healthcare delivery ecosystems to the automotive marketplace to urban ocean systems.

Visitors to the lab often find these large interactive visualizations both compelling and pleasing. Starting in 2015, we began a series of investigations to determine whether these interactive visualizations actually improve decision making. This was motivated in part by many comments from industry partners that they too provided such visualizations and their customers loved them. However, no one had any empirical evidence that better decisions resulted.

Government Sponsors

I address the two largest government sponsors of academic research—NIH and NSF—later in this chapter. In this section, I relate a few experiences of dealing with government sponsors.

In 1975 while at Illinois, I developed a proposal for a computational model of statewide delivery of information services. It was an ambitious project with a modest budget. I met with the state official who could decide to fund the project. He and his staff listened to my pitch and asked a few questions about how the model could provide value to them. Then, the official looked at the budget and said, "I like your proposal. I'll give you half of your requested budget. Take it or leave it." I thought about it quickly, took the offer, and received bigger budgets over the next couple of years.

During the annual review of Joint Services Electronics Program at Illinois in 1976, presenters were allowed one slide and 5 minutes of presentation time per $15,000 of funding. This meant that I was allowed one slide! It was a struggle to provide any detail about my project without making the slide unreadable. Then, I realized that I could accompany my one slide with 5 minutes of elaboration of things not on the slide. This changed my view of slides. They are just reminders of the story you want to tell. So my one slide was just a few key phrases and my story was much richer. My project was renewed for another year.

In 1981, now at Georgia Tech, I visited the NSF program manager overseeing my grant and many others. I wanted to report to her what I felt were impressive experimental results from our study of human search behavior using computer-based information retrieval. I provided such in-person reports to my other sponsors at the Army, Navy, and NASA. When I told her why I came to her office, she asked, "What if everybody did this?" She told me that she could not consume the level of detail I wanted to provide from her large number of grantees.

Industry Sponsors

I have had many industry research sponsors. At Georgia Tech, my industry sponsors included Lockheed Martin, IBM, General Motors, and Dollar General. Stevens' industry sponsors of my research include Accenture, Lockheed Martin, IBM, and Northern Light. I recently had an opportunity to discuss with United Technologies their sponsorship of a research institute at the University of Connecticut.

All of these companies share a few opinions in particular. They feel that faculty members and graduate students are very talented. Further, they feel that the cost of university research is quite reasonable. However, they do not feel that academia is willing to work at the pace industry needs. One executive told me, "The unit of time in academia is one semester. When a faculty member tells you that they will do something right away, they mean by the end of the semester."

RESEARCH MODEL

Research involves submitting proposals, conducting the proposed research, and writing articles and submitting them to journals. In this section, I will first address the probability of success when submitting an article. Then I will consider the expected accumulations of citations subsequent to an article being published. Next I will discuss how to project the probability of success in getting a research proposal funded. Finally, a particular scenario is used to illustrate a key trade-off.

Submission of Articles

The probability of an article being accepted for publication is given by Equation 9.1:

$$\text{PA} = P_{OA} \exp(-\lambda_A \text{NST}) \quad (9.1)$$

where PA is the probability of acceptance, NST is the total number of articles submitted by all authors, and P_{OA} and λ_A are model parameters fit to data for different journals.

Data were collected for the journal *Nature* (Nature, 2015), a compilation of fifty *IEEE Transactions* (Zappula, 2015), and the relatively new *Journal of Systems Engineering* (de Weck, 2015). The best-fit parameters (minimal root mean-squared error) for Equation 9.1 for these three journals are shown in Table 9.1, along with the annual growth rate for NST.

Figures 9.1, 9.2, and 9.3 show comparisons of data and models for the three journals. Clearly, it is most difficult to have articles accepted by *Nature*. The chances are better with *IEEE Transactions* and better yet with *Systems Engineering*. With a larger λ_A, *Systems Engineering* will soon have an acceptance rate that is comparable to *IEEE Transactions*. Both of these journals have very high growth rates for submissions that, following Equation 9.1, will decrease PA over time. Of course, as IEEE has repeatedly demonstrated, increasing demand can prompt establishing new journals to accommodate the demand.

If a faculty member submits NS articles, at a cost of a percent of their time per article, then the number accepted NA will be the product of PA and NS. It is assumed

TABLE 9.1 Model Parameters for Three Journals

Journal	P_{OA}	λ_A	NST Growth (%)
Nature	24	0.0001	2.25
IEEE Transactions	56	0.00001	9.0
Systems Engineering	86	0.01	8.0

RESEARCH MODEL

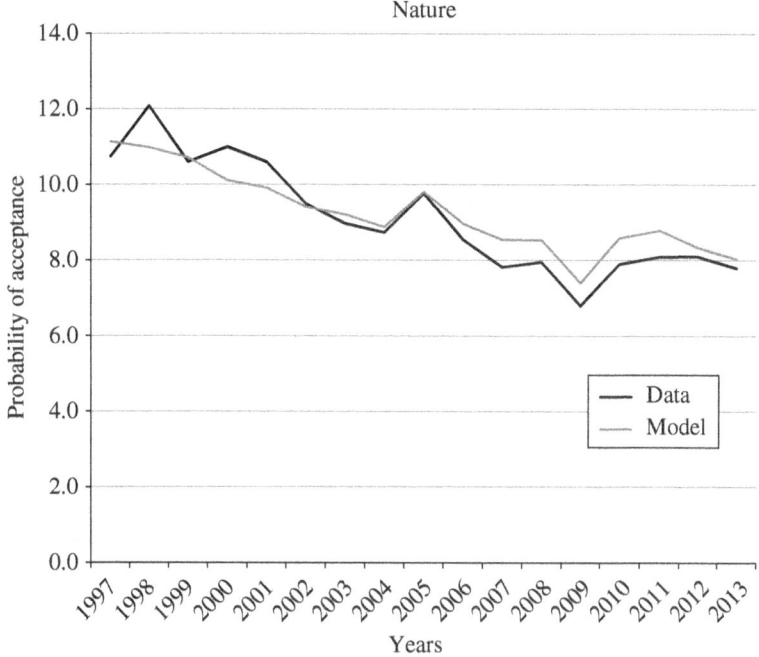

FIGURE 9.1 Data and model for *Nature*.

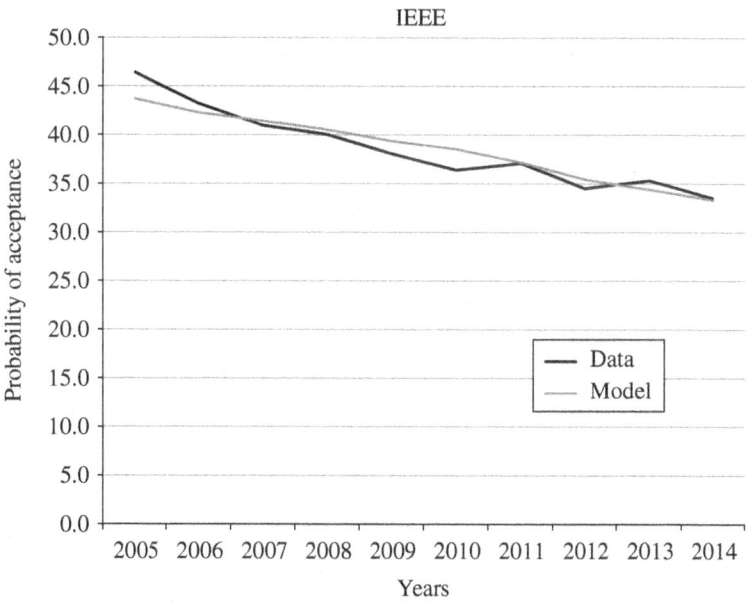

FIGURE 9.2 Data and model for *IEEE Transactions*.

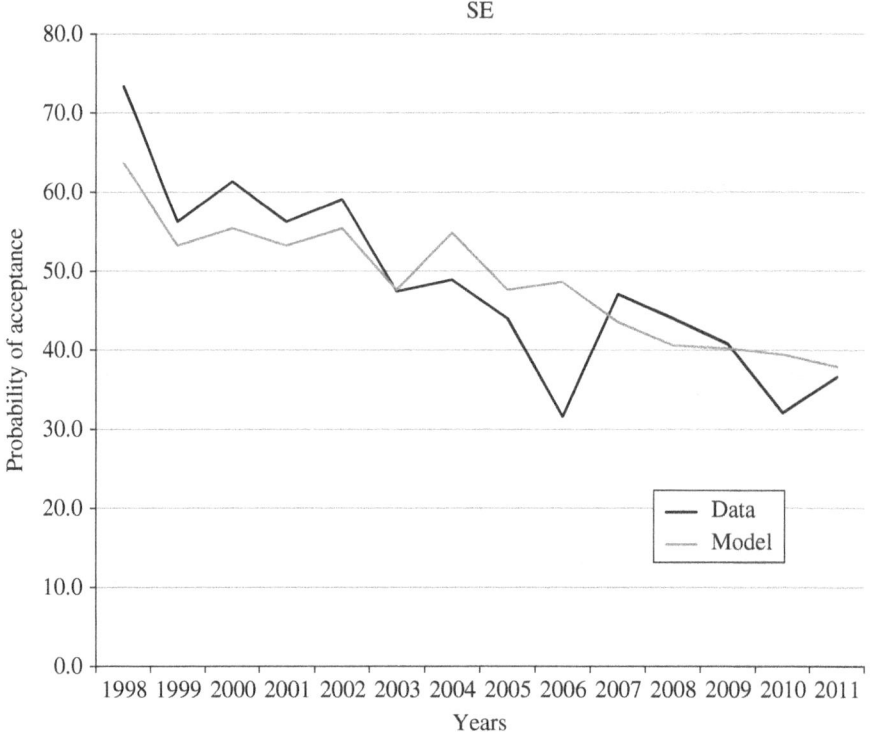

FIGURE 9.3 Data and model for *Systems Engineering*.

that a faculty member will use all their research time preparing articles other than the time spent preparing proposals. This assumption is, of course, easy to vary.

Citation of Articles

Cumulative citations are modeled by Equation 9.2:

$$\mathrm{NC}(T) = \mathrm{NC}_0 \left[1 - \exp(-\lambda_c T)\right] \qquad (9.2)$$

where NC (T) is the cumulative number of citations T years after publication and NC_0 and λ_c are model parameters fit to citation patterns for different disciplines. For patterns averaged across all science and technology disciplines, the best-fit parameters (minimal root mean-squared error) for Equation 9.2 are NC_0 equals 24 and λ_c equals 0.125. Data are available for each discipline if needed (THE, 2011).

Cumulative citations, over time (Figure 9.4), can be used to compute a faculty member's h-index, denoted by HI. This index is defined as the number of published articles that have at least HI citations. For example, HI equal to 20 means that there are 20 articles with at least 20 citations. Article number 21 has, by definition, less than 21 citations. Otherwise, HI would be 21.

RESEARCH MODEL

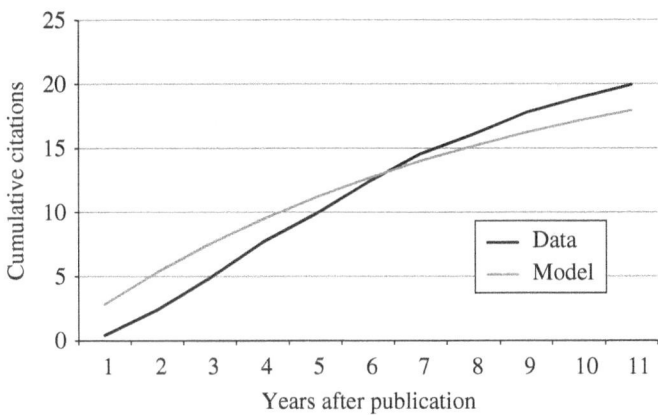

FIGURE 9.4 Data and model for cumulative citations.

TABLE 9.2 Model Parameters for Two Agencies

Agency	P_{OF}	λ_F	NPT Growth (%)
National Institutes of Health	81	0.000024	4.5
National Science Foundation	76	0.000025	3.6

Submission of Proposals

The probability of a proposal being funded is modeled by Equation 9.3:

$$\text{PF} = P_{OF}\exp(-\lambda_F \text{NPT}) \tag{9.3}$$

where PF is the probability of funding, NPT is the total number of proposals submitted by all researchers, and P_{OF} and λ_F are model parameters fit to data for different funding sources.

Data were collected for the National Institutes of Health (NIH, 2015) and the National Science Foundation (NSB, 2014). The best-fit parameters (minimal root mean-squared error) for Equation 9.3 for these two agencies are shown in Table 9.2, along with the annual growth rate for NPT.

Figures 9.5 and 9.6 show comparisons of data and models for the two agencies. Note that the NSF data for 2009 and 2010 include funds made available under the American Recovery and Reinvestment Act of 2009. The model fit degrades for 2009, but this is irrelevant for future years.

It appears that the chances of success are a bit higher for NSF than NIH. However, the percent funded varies by division, for example, the Engineering Division was as low as 13% in 2005 (IPAMM, 2007). The percent funded varies significantly with award mechanisms. The median annualized award increased from $85,000 in 2002 to $130,000 in 2013. The mean award duration remained at roughly 3 years throughout.

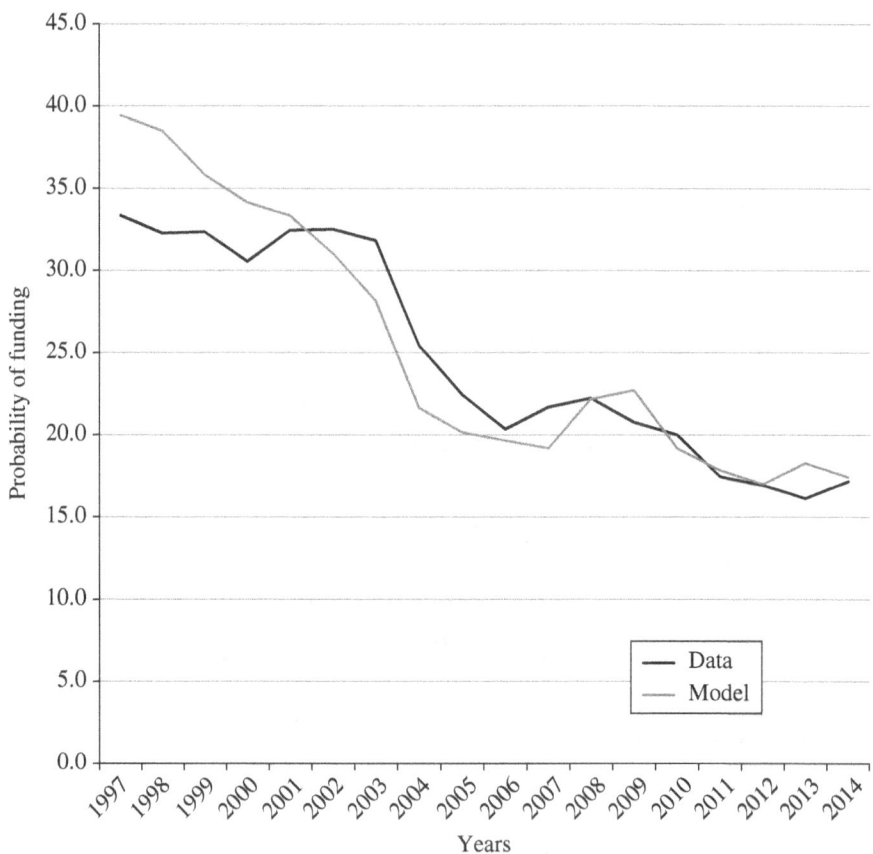

FIGURE 9.5 Data and model for National Institutes of Health.

NSF will typically not pay any portion of a faculty member's salary during the academic year. However, they do allow summer salary. The average number of months of salary paid decreased from 1.96 in 2002 to 0.81 in 2013. The average number of months of salary paid for proposals funded by the Engineering Directorate deceased from 1.1 to 0.4. Thus, if NSF grants were a faculty member's only source of funding, he or she would need three to six funded projects to have a full summer salary.

If a faculty member submits NP proposals, at a cost of a percent of their time per proposal, then the number funded NF will be the product of PF and NP. It is assumed that a faculty member will submit the number of proposals that will assure an expected grant every other year. This assumption is also easy to vary.

Overall Model

The model outlined here enables predicting NA, NC, and NF over time for each faculty member. Putting the pieces together, assume that a faculty member only submits proposals to NIH and only submits articles to *Nature*. This person is tenure track and,

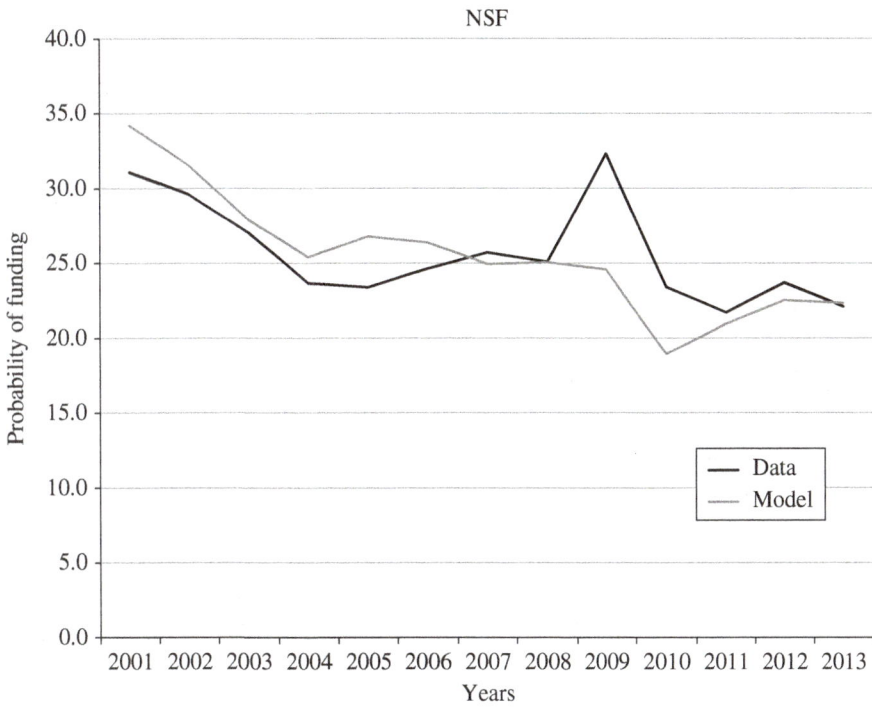

FIGURE 9.6 Data and model for National Science Foundation.

therefore, can devote half of their time to research. Further, assume that it takes one person-month to prepare a proposal and one person-month to write an article. Finally, assume that the faculty member's priorities are such that he or she wants to assure winning a grant award every other year. Thus, they prepare the number of proposals needed to yield an expected award rate of 0.5 per year. They devote the reminder of their research time to writing and submitting articles.

Figure 9.7 shows the projected results of this strategy. Over 20 years, the number of proposals submitted (NP) increases from 1.5 to 3.5 per year. The number of articles submitted (NS) decreases from 4.7 to 2.5 per year. The number of articles accepted (NA) decreases from 0.4 to 0.1. Overall, five papers are published over 20 years. Fortunately, this faculty member's early productivity might have earned them tenure after the sixth year, but steadily declining productivity would be very unlikely to get them promoted to full professor by the 10th or 12th year. The strategy of only seeking funding from a highly esteemed sponsor and only publishing in a highly prestigious journal completely undermines this faculty member's career.

Figure 9.8 shows the results when the hypothetical faculty member chooses to submit articles to *IEEE Transactions* rather than *Nature*. Not surprisingly, the results significantly improve, with 15 papers published over 20 years. However, this faculty member is still plagued by the steadily declining likelihood of success at NIH. He or she would have to decrease the time spent on NIH proposals to focus on writing and submitting the articles that would get them promoted to full professor.

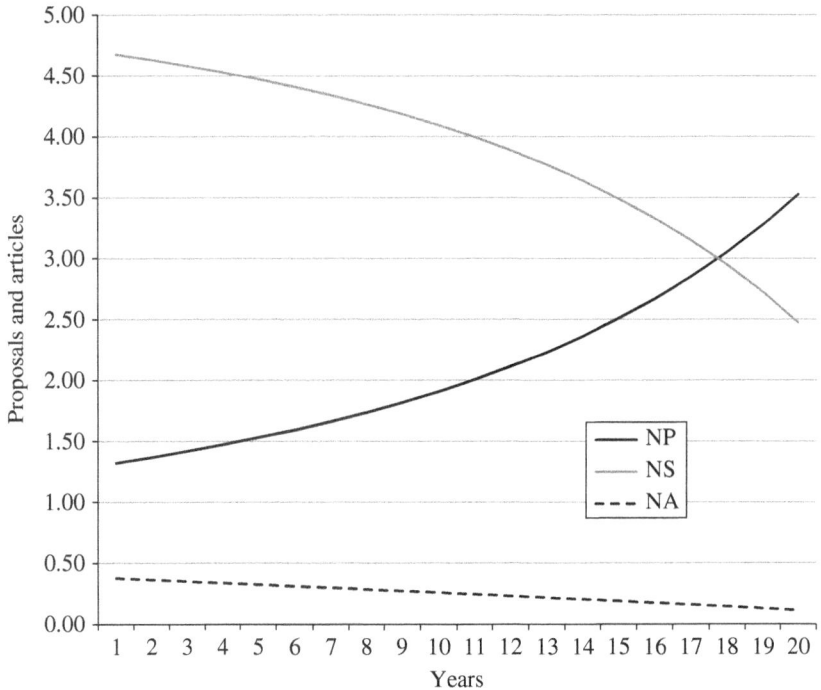

FIGURE 9.7 Proposals submitted and articles submitted and accepted.

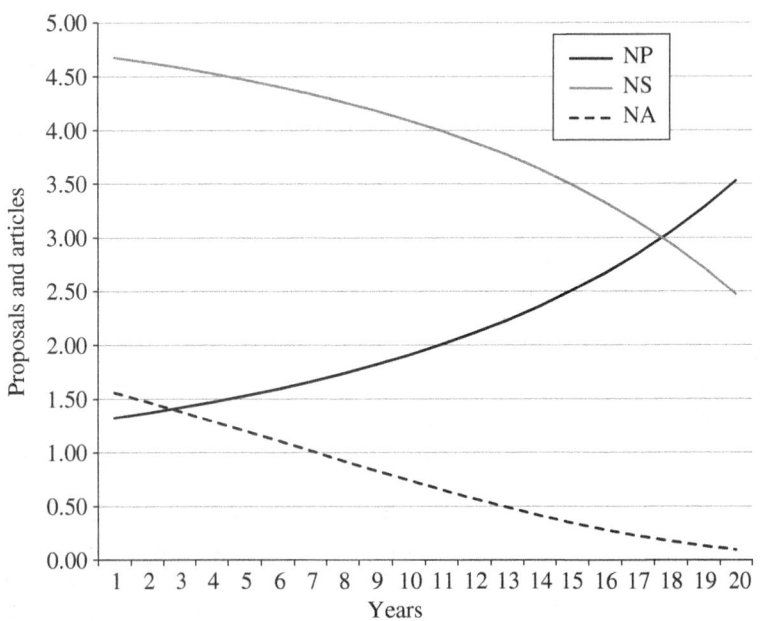

FIGURE 9.8 Proposals submitted and articles submitted and accepted.

Chapter 7 presented a tenure model focused on the probabilistic nature of tenure decisions. Chapter 8 presented a workforce model that incorporates tenure decisions. This chapter presented a research model that generates the publications, citations, and funding needed to earn tenure. In Chapter 10, a brand value model is presented, the predictions from which serve as a proxy for rankings. All of these models, as well as elements from other chapters, come together as an integrated model in Chapter 12.

INTELLECTUAL PROPERTY

The Goldwater–Nichols Act empowers universities to commercially exploit the results of federally funded research. Corporately funded research often results in jointly owned IP. Universities have come to increasingly control IP, although they have a poor track record of exploiting it.

Gatorade has provided enormous income to the University of Florida and Genentech has been a windfall for the University of California at San Francisco. However, the number of major successes like these is dwarfed by the vast numbers of ideas that never lead to any revenues and profits for the licensee or royalties for the university. Perez-Pena (2013) reports that patenting and licensing are seldom profitable. Revenues generated are seldom greater than the costs of running the IP patenting and licensing function. Further, rigid IP policies tend to alienate sponsors.

In some cases, IP takes the form of patents, but more often it is embodied in the know-how of the researchers. Either form of IP may result in spin-off businesses that create jobs, revenue, profit, and perhaps public stock offerings. As discussed in Chapter 3, spin-offs enabled MIT research, by 1997, to foster the formation of 4000 companies employing 1.1 million employees with sales of $232 billion.

Many universities require faculty members, and sometimes students, to sign agreements that indicate the university owns any IP created. One university forces faculty to agree to the university owning any know-how gained while in the university's employ. One IP lawyer told me that owning the know-how of others is completely unenforceable.

Faculty members are usually allowed to consult 1 day per week and any IP created for consulting clients are not subject to such university agreements. Faculty who spin off technologies to launch new business ventures are usually encouraged, but the "rules of this game" vary greatly, with differing levels of incentives and impediments.

Spin-Off Experiences

In 1980, while at Illinois, the employer of a former graduate student called to ask if we would like to commercialize simulation-oriented computer-based instruction, a line of research that we had been pursuing for several years. The caller was the president of Marine Safety International and he was interested in creating desktop simulators for marine engineers to learn about supertanker propulsion systems prior to being immersed in the company's multimillion dollar full-scope engine room simulator.

This was serendipity at its best. I had just returned from a year at Delft University of Technology where, among other things, I had studied the problem-solving skills of supertanker engineers. We were ready for this opportunity. We formed Search Technology, Inc. in late 1980. For the next decade, the company sold simulation-oriented computer-based instruction packages to train operators and maintainers of a wide range of systems.

Lockheed Martin contacted us in 1984. They had happened upon several of my articles related to human interaction with intelligent systems. They wanted to interest us in becoming a subcontractor on a proposal effort to compete for DARPA's Pilot's Associate contract. We won. It resulted in many millions of dollars of revenue. A couple of years later, we won the Air Force's Designer's Associate contract, which was somewhat smaller but still several million dollars. We soon had almost 40 employees.

In the mid-1980s, one of our senior staff members had an idea for a very low-cost flight management system driven by what we would now call advanced analytics. We funded the founding of Search Aeronautics to take this idea to market. We made one large sale to Air Canada. The company was then sold to the staff member and his partner for a quick and reasonable profit.

Customers often asked about our approach to designing training and aiding systems; the former creates the potential to perform while the latter directly augments performance. They asked us to teach them what we came to call human-centered design. We began to offer short courses. Customers for these human-centered design short courses asked if we would write books that integrated the course materials. Three volumes resulted, all published by John Wiley.

In the early 1990s, customers for our human-centered product planning workshops and consulting asked if we could help them rethink their overall business processes. This led to our focusing on the overall enterprise, not just new product R&D. As this portion of the business grew, we formed a wholly owned subsidiary, Enterprise Support Systems (ESS). I later traded my shares in Search Technology to acquire all the shares of ESS and it became independent in 1995.

Our customers for planning workshops and consulting services asked if we could develop computer-based tools that embodied the three books. This led to the *Advisor Series* suite of four tools:

1. *Product Planning Advisor*
2. *Strategic Planning Advisor*
3. *Situation Assessment Advisor*
4. *Technology Investment Advisor*

Contrary to our expectations, software sales never exceeded 20–30% of our sales, although software accounted for a much higher percentage of our profits. Customers demanded consulting services to support use of our tools. They also requested that we undertake a wider range of projects.

In 2001, I returned to Georgia Tech as chair of the School of Industrial and Systems Engineering. My plan had long been to return to academia and the timing of this opportunity was good, both professionally and economically. My 13 years of running two software companies and, in particular, my intense interactions with executives in over 100 companies resulted in my research portfolio morphing to focus on enterprises as systems and especially enterprise transformation. This led to the founding of the Tennenbaum Institute at Georgia Tech and, more recently, the Center for Complex Systems and Enterprises at Stevens.

I am currently planning a new venture, which should be formally launched by the time this book appears in print. We intend to spin off our healthcare delivery models and simulations in the form of "software as a service." We expect that, as with ESS, we will need to include consulting service offerings to adapt our models and simulations to particular customers. The company is named Curis Meditor, which is Latin for healthcare best practices.

CONCLUSIONS

This chapter addressed the challenges of research in terms of the role of peer review of publications and proposals, bibliometrics associated with citations, and difficulties securing funding. A variety of research experiences were related. A model of research was presented that predicts probabilities of articles being accepted, expected citations after publication, and probabilities of research proposals being funded. This model is a central element of the integrated model discussed in Chapter 12.

The last major section of this chapter addressed IP, which is often one of the outcomes of research. In some cases, this takes the form of patents, but more often it is embodied in the know-how of the researchers. Either form of IP may result in spin-off businesses that create jobs, revenue, profit, and perhaps public stock offerings. I related several of my experiences in launching and growing spin-offs.

REFERENCES

Bornmann, L., & Mutz, R. (2015). Growth rates of modern science: A bibliometric analysis based on the number of publications and cited references. *Journal of the Association for Information Science and Technology, 66* (11), 2215–2222.

de Weck, O. (2015). Personal communication. June 21.

Editors (2011). Dr. No Money: The broken science funding system: Scientists spend too much time raising cash instead of doing experiments. *Scientific American*, April 19.

Harley, D., & Acord, S.K. (2011). *Peer Review in Academic Promotion and Publishing: Its Meaning, Locus, and Future.* Berkeley, CA: Center for Studies in Higher Education, University of California at Berkeley.

Howard, D.J., & Laird, F.N. (2013). The new normal if funding university science. *Issues in Science and Technology, 30* (1).

Ioannidis, J.P.A. (2014). Is your most cited work your best? *Nature, 514*, October 30.

IPAMM (2007). *Discussions with the Advisory Committee.* Washington, DC: NSF Working Group on the Impact of Proposal and Award Management Mechanisms.

Morse, P.M. (1968). *Library Effectiveness: A Systems Approach.* Cambridge, MA: MIT Press.

Narin, F. (1976). *Evaluative Bibliometrics: The Use of Publication and Citation Analysis in the Evaluation of Scientific Activity.* Cherry Hill, NJ: Computer Horizons.

Nature (2015). http://www.nature.com/nature/authors/get_published/ (accessed February 8, 2016).

NIH (2015). Source: http://www.report.nih.gov/success_rates/index.aspx (accessed August 1, 2015).

NSB (2014). *Report to the National Science Board on the National Science Foundation's Merit Review Process (NSB 14–32).* Washington, DC: National Science Foundation.

Perez-Pena, R. (2013). Patenting their discoveries does not payoff for most universities. New York Times, November 20.

Pritchard, A. (1969). Statistical bibliography or bibliometrics. *Journal of Documentation, 25* (4), 348–349.

Rouse, W.B., & Serban, N. (2014). *Understanding and Managing the Complexity of Healthcare.* Cambridge, MA: MIT Press.

Saracevic, T. (2004). *Bibliometrics.* New Brunswick, NJ: School of Communication and Information, Rutgers University.

THE (2011). Citation averages, 2000–2010, by fields and years. *Times Higher Education*, May 31.

Thomson-Reuters (2008). *Using Bibliometrics: A Guide to Evaluating Research Performance with Citation Data.* London, UK: Thomson-Reuters.

Uzoka, F.-M., Fedoruk, A., Osakwe, A., Osuji, J., & Gibb, K. (2013). A multi-criteria framework for assessing scholarship based on Boyer's scholarship model. *Information Knowledge Systems Management, 12* (1), 25–51.

Ware, M., & Mabe, M. (2012). *An Overview of Scientific and Scholarly Journal Publishing.* The Hague, Netherlands: International Association of Scientific, Technical, and Medical Publishers.

Werner, R. (2015). The focus on bibliometrics makes papers less useful. *Nature, 517*, January 15.

Zappula, F. (2015). Personal communication. September 3.

10

RANKINGS AND BRAND VALUE

The globalization of university-based science and engineering education and research is associated with the creation of national and international "brands" by leading research universities. Such branding is reflected in rankings of universities and their programs. High brand visibility appears to lead to high rankings and vice versa. This chapter explores this phenomenon for university-based science and engineering programs.

This chapter addresses several ranking schemes that emphasize different attributes of a university. Georgia Tech is used as an example to illustrate the types of initiatives needed to improve rankings. Statistical analyses of rankings of one program—industrial/manufacturing/systems engineering—are reported. The best predictors of a program's ranking are last year's ranking plus the number of faculty.

The brand value (BV) of a university is discussed as a proxy for rankings. An index of BV is presented that is a weighted sum of number of articles published, number of citations received, and h-index, totaled across all faculty members of an institution. An example is used to illustrate the nuances of the BV index as affected by research sponsors and publication outlets chosen by faculty members.

Universities as Complex Enterprises: How Academia Works, Why It Works These Ways, and Where the University Enterprise Is Headed, First Edition. William B. Rouse.
© 2016 John Wiley & Sons, Inc. Published 2016 by John Wiley & Sons, Inc.

RANKING SCHEMES

There are several ranking systems that receive much attention. The annual *USN&WR* (2003) system was started in1983 and has become a key source for high school juniors deciding where to apply for college and college juniors and graduates deciding where to apply for graduate study. The National Research Council (NRC) (1995) performs more in-depth evaluations, roughly every 10 years. While academics tend to give more credence to the NRC rankings, the general public is much more aware of the *USN&WR* rankings. I also discuss the Shanghai Jiao Tong University and *Times Higher Education* ranking schemes.

For establishing the rankings of undergraduate and graduate engineering programs, there are two major views offered by *USN&WR*—one ranking for overall university programs, for example, engineering, and other rankings for the different disciplines within these programs. In addition, *USN&WR* differentiates between those schools whose terminal degrees are masters versus doctoral. For the overall program ranks, several variables are measured and the means are standardized, scaled, and weighted so as to produce an overall score. The most recent weightings for undergraduate programs are:

- Undergraduate academic reputation, 22.5%
- Graduation and freshman retention rates, 20%
- Faculty resources, 20%
- Student selectivity, 15%
- Financial resources, 10%
- Graduation rate performance, 7.5%
- Alumni giving, 5%

Rankings are based on these scores. For the individual programs, the ranks are based solely on judgments of deans and chairs. For each of the schools surveyed in each discipline, the deans and chairs are asked to rate them according to the level of excellence they have achieved in the particular disciplines. The votes are tallied and the ranks are reported.

The disciplines surveyed in the engineering field are:

- Aerospace/aeronautical/astronautical
- Bioengineering/biomedical
- Chemical
- Civil
- Computer
- Electrical/electronic/communications
- Environmental/environmental health
- Industrial/manufacturing/systems
- Materials

- Mechanical
- Nuclear
- Petroleum

As discussed later, Georgia Tech, for many years, has been ranked in the top 5 overall and in the top 10 for at least 7 of the 12 disciplinary categories. Specifically, as discussed more in depth in the following, Georgia Tech has achieved the number 1 rank in the industrial/manufacturing/systems engineering category for over 20 years. This has been a tremendous change in the status quo, and the story of this change can provide insights to other institutions as to how to change their ranks and reputations in a domain that seems to be characterized by much inertia.

The 1995 NRC report included its findings from a 4-year study on the rankings of research-doctorate programs in the United States. The study was conducted as an update of the first such publication in 1982. The NRC conducted both reputational surveys and rankings of objective characteristics in order to arrive at compound measures of university rankings. Different disciplines were researched, providing separate ranks for humanities, sciences, and engineering and the subdisciplines within these broader categories. Based on both the reputational and objective characteristic surveys and ratings, Georgia Tech's Industrial and Systems Engineering program was ranked number 1 in this report. This provides some credence to the annual *USN&WR* rankings but is, of course, far from a definitive assessment of the parallels between the two sources.

Much more recent are the rankings by Shanghai Jiao Tong University and *Times Higher Education* (Marginson, 2007). For ranking of undergraduate programs, *USN&WR* focuses on aspects of institutions seen to contribute to the quality of teaching and the student experience rather than research and scholarship. In contrast, the Shanghai Jiao Tong rankings focus national government attention on policies designed to concentrate research activity in a small number of universities. Higher education is seen as about scientific research and Nobel Prizes. Their attributes and weightings are:

- Alumni of institution: Nobel Prizes and field medals, 10%
- Staff of institution: Nobel Prizes and field medals, 20%
- High citation researchers, 20%
- Articles in citation indices in sciences, social sciences, and humanities, 20%
- Articles in *Science* and *Nature*, 20%
- Research performance per staff member, 10%

The *Times Higher Education* ranking sees higher education as primarily about building reputation as an end in itself, and about international marketing, because it is these metrics that drive their index. Their attributes and weightings are:

- Peer review survey of academics, 40%
- Survey of "global employers," 10%

- Proportion of academic faculty who are foreign, 5%
- Proportion of students who are foreign, 5%
- Staff–student ratio (proxy for teaching quality), 20%
- Research citations per head of academic faculty, 20%

The various ranking schemes are not without critics. Gladwell (2011) argues that the *USN&WR* rankings suffer from there being "no direct way to measure the quality of an institution—how well a college manages to inform, inspire, and challenge its students. So the U.S. News algorithm relies instead on proxies for quality—and the proxies for educational quality turn out to be flimsy at best."

He notes that "According to educational researchers, arguably the most important variable in a successful college education is a vague but crucial concept called student 'engagement'—that is, the extent to which students immerse themselves in the intellectual and social life of their college—and a major component of engagement is the quality of a student's contacts with faculty." He asserts that the proxy measures employed by *USN&WR* for this phenomenon are very weak choices.

Quiggin (2015) reports that of the 20 top universities in 2015, 16 were among the top 20 in 1911. In contrast, of the companies in the Dow Jones Industrial Average in 2015, only one was in this index in 1911—General Electric. This provides credence to the belief that change in academia is indeed very slow. An example of much more rapid change is discussed in the next section

Despite the criticisms, university rankings have become increasingly important to administrators as students and parents place greater emphasis on rankings for selecting where to apply and enroll. Ruhvargers (2012) presents a history of ranking schemes and elaborates the sometimes arbitrary scaling and weighting of factors in these schemes. Clarke (2002) presents guidelines for choosing among and designing guidelines for quality ratings.

EXAMPLE OF MOVING UP

The Georgia Institute of Technology was founded in 1885. Up until the early 1970s, Tech's reputation was as an excellent undergraduate engineering school. Larger aspirations emerged with the presidency of Joseph Pettit (1972–1987), who arrived from having served as Dean of Engineering at Stanford University. Pettit's emphasis on Ph.D. research began the Institute's remarkable climb from being ranked a top 20 engineering program in the 1980s to top 10 in the early 1990s and top 5 since 1997. As later analyses indicate, such a rapid improvement is indeed quite an accomplishment.

During the recent years, enrollment has grown only modestly, slowing shifting the balance toward graduate education. The number of faculty has also only grown modestly. However, almost 80% of the current faculty members were hired in the past 10–20 years. This is an amazing level of turnover, especially given the tenure system in academia.

Almost 5% of faculty members have been elected to the prestigious national academies. Roughly one-eighth hold endowed chairs or professorships. Over one-eighth have won coveted career awards from the National Science Foundation. Thus, the turnover has resulted in greatly increased excellence among the faculty, in addition to infusing the university with new and fresh perspectives.

During this time, annual awards of research grants and contracts have doubled, as has the Institute's overall budget. The percentage of the budget coming from the State of Georgia has continually declined, currently at roughly 15%. Decreasing state support of public institutions is a nationwide phenomenon, and all research universities have actively pursued several other funding sources so as not to suffer as a result of reduced state university budgets.

The quality of incoming students has continued to rise during this period. Average scores on Scholastic Achievement Tests are above 1400 on a 1600-point scale. The mean high school grade point average is 3.80 on a 4.0 scale. Undergraduate degrees now account for only 60% of degrees granted.

The changes initiated by Pettit and enhanced by his successors, John P. Crecine (1987–1994) and Wayne Clough (1995–2008), can be summarized as follows:

- Greatly increased emphasis on Ph.D. programs and sponsored research:
  - Plus increased emphasis on multidisciplinary research and education
  - Plus substantial increase of endowment, for example, faculty chairs
  - Plus substantial expansion and upgrade of research and education facilities
  - Plus increased emphasis on university's role in economic development
- Top-down vision and leadership with bottom-up strategy and execution:
  - With strong entrepreneurial institutional culture

Also of note is the 1996 Olympics hosted by Atlanta. This resulted in substantial investments in the Institute's infrastructure, ranging from new dormitories and athletic venues to greatly enhanced landscaping across campus. In parallel, a capital campaign targeted to raise $300 million during this period yielded almost $800 million and safely concluded before the Internet "bubble" burst. One of the primary uses of these resources was a dramatic increase in the number of chaired positions, thereby enabling the attraction of the "best and brightest."

DETERMINANTS OF RANKINGS

It is natural to wonder what actually affects the perceptions that, in turn, affect rankings of educational programs. The influences just summarized for Georgia Tech represent a consensus of current and past leaders of the Institute. However, these conclusions are far from scientifically rigorous. What do we really know about the determinants of rankings?

This question caused us to explore in depth the available data for one field—industrial engineering and manufacturing (Rouse & Garcia, 2004). The School of Industrial and Systems Engineering (ISyE) at Georgia Tech has enjoyed the top ranking in this field for many years, in fact for all the years that these rankings have been reported, save one, over 20 years ago.[1]

As chair of this school, I was interested in why the *U.S. News & World Report* awards ISyE this ranking, as well as what we should do to preserve this position. Based on the different methods of ranking overall university programs and disciplines within these programs, it is clear that the latter are rather subjective. Therefore, we wanted to explore whether the subjective rankings could be predicted by or correlated with more quantitative, objective measures. We had the good fortune to be able to address this question in some depth due to the availability of data from a long-standing annual benchmarking study performed among all the leading programs in this field.

The a priori expectation is that certain variables, such as those that measure size of school and research dollars, will be correlated and/or able to predict changes in rankings, after a certain time period. Deans are generally aware of major changes within their peer group of schools, and although it may take a few years for the changes to reflect in the rankings, we believed that there should be some predictability in the changes of school rankings based on previous changes in benchmark data.

Each year, we emailed all the top programs a spreadsheet that includes entries for the following items:

- Degrees awarded—number and type
- Enrollment—for undergraduate, masters, and doctoral programs
- Student information—SAT scores, GRE scores, etc.
- Teaching loads
- Sections taught
- Number of faculty—full, associate, and assistant professors
- Faculty honors and awards—National Academies membership, Society Fellows, etc.
- Faculty salaries
- Research support
- Staff
- Space

There are also a variety of subitems within each of the categories. These data were available for all the top-ranked programs for roughly the same 15-year period for which we had rankings data.

---

[1] When I became chair of ISyE in 2001, we had been ranked no. 1 for over 10 years. A very influential ISyE alum said to me that he understood that the no. 1 ranking was not all that matters, but he counseled, "Don't lose it!" I didn't.

Statistical analysis of these data via time series analysis and correlation matrices indicated several interesting conclusions. First, over 95% of the variance in any university's ranking is explained by their ranking the previous year. Thus, from year to year, there is a very high level of inertia in the system. This suggests that Georgia Tech's climb during the past 10–15 years is truly unusual.

Second, over many years, the best predictors of a program's ranking are:

- Number of faculty—rank-order correlation of 0.3–0.5 with a typical lag of 4–5 years
- Number of graduate degrees awarded—rank-order correlation of 0.5–0.6 with a typical lag of 1–2 years
- Number of undergraduate degrees awarded—rank-order correlation of 0.5–0.6 with a typical lag of 1–2 years

Clearly, size matters.[2] Possible interpretations of this conclusion are elaborated in the following.

Third, contrary to the NRC findings, level of research support per faculty member was negatively correlated, albeit weakly (0.1–0.2), with a university's ranking, particularly for the less recent years. This was due to two lower-ranked programs—but still top 10—having once had much higher levels of support than the other highly ranked programs. This illustrates the problems of small data sets—15 years of data for 10 universities.

Interestingly, our data set was rather unusual. These types of data are seldom available for such a long period of time. In other words, this data set was about as good as it gets but still insufficient to avoid possibly anomalous results. Of course, lack of data has not deterred various pundits from articulating "truths" about rankings. Our experience is that expertise in one particular science or engineering discipline appears to enable reaching conclusions about social and organizational phenomena without data.

Returning to the issue of size, I developed several hypotheses about the underlying phenomena. Faculty size, for example, predicts rankings because, I think, having more faculty members increases the likelihood of having more well-known "stars." These stars, I hypothesize, provide the impetus for higher rankings rather than the simple number of faculty members, especially given the subjective and reputational bases of these rankings.

Similarly, the number of graduates is not the underlying predictor. My hypothesis is that the larger number of graduates increases the probability of outstanding leaders in academia, industry, and government who provide perceived evidence of the excellence of the university. Size matters because it provides more opportunities for excellence.

---

[2] With one exception, the NRC report found similar correlations and associations as did we in our analyses of benchmark versus rank data. Both size—as measured by the number of faculty and graduate students—and involvement in research were found to be highly correlated with the quality assessment and rank of a program.

In other words, size provides more opportunities for "tipping points" where big differences suddenly happen (Gladwell, 2000). Hoards of faculty and graduates are not the driver of rankings. However, these hoards provide increased opportunities for the people who make big differences in the world to be associated with your university.

Of course, talent also counts. So, what you want is hoards of very talented people. This almost guarantees excellence. The question then becomes one of how to get the most talent into your university. MIT, for example, has known how to do this for 50 years or more. Georgia Tech has figured this out over the past 20 years. Given that talent is not unlimited, the determining competency is the ability to get the talent to choose you.

As discussed in Chapter 8, this is also true in terms of attracting talented students. I recall one late evening when I was a Ph.D. student at MIT in the early 1970s. I was working with a group of Ph.D. students trying to solve a thorny homework problem on a hybrid computer, that is, an integration of analog and digital computers. We finally succeeded around midnight. One of the students commented, "We taught each other a lot tonight. I am glad MIT attracts such talented classmates."

BRAND VALUE

A university's BV plays an important role in several phenomena. As noted in the following, it affects students' choices of where to apply and, if accepted, enroll. Hoyt and Brown (2003) and Noel-Levitz (2012) discuss many other factors that affect students' choices, but BV and costs dominate.

BV also affects alumni relations with the university (Alexander et al., 2006; Svoboda & Harantova, 2014). Alumni who feel a strong relationship with their alma mater's brand are more likely to encourage their children to attend, support various programs, and perhaps provide gifts and endowments to advance the vision and mission of the university.

BV, as elaborated in the following, is a proxy for rankings. The ranking schemes discussed earlier include a variety of factors that are difficult to predict. Instead, we would like to use objective and inherently measurable attributes of a university that, when combined into a BV index, result in rank orderings of research universities that are highly correlated with the rank orderings from the ranking schemes reviewed earlier.

Model of Brand Value

Figure 10.1 (repeated from Chapter 6) depicts students' choices of where to apply and enroll as being most influenced by tuition (net of financial aid) and BV (Hoyt & Brown, 2003; Noel-Levitz, 2012). Tuition is an input parameter on the overall dashboard discussed in Chapter 12. We need to project BV.

It is easy to imagine an enormous number of factors associated with quality of faculty, students, and infrastructure that influence perceptions of BV. Fortunately,

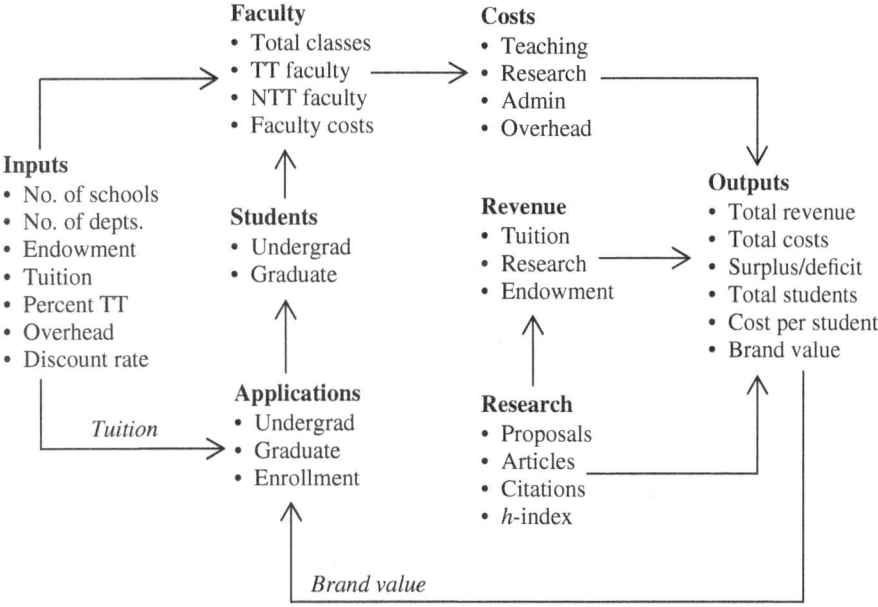

FIGURE 10.1 Overall structure of economic model of academic enterprises.

Lombardi and colleagues (2012) report the results of a very ambitious and interesting analysis. They compared rankings of institutions published by various sources with rankings based on a simple formula, shown in Equation 10.1:

$$BV(t) = \alpha NA(t) + \beta NC(t) + \delta HI(t) \qquad (10.1)$$

where BV (*t*) equals the BV at time *t* and the inputs NA, NC, and HI are totaled across an institution's faculty members, drawing upon the Global Research Benchmarking System (GRBS) of the United Nations University International Institute for Software Technology. They found that ranking research universities by BV produced rank orders that were very similar to those created by much more elaborate schemes.

Within the overall model used in Chapter 12, the coefficients were all set equal to 0.333. Further, I divide NA by 10 and NC by 100. This is done to assure that none of these factors, particularly NC, dominate the BV projections. Since BV is used as a measure to compare policies or scenarios, the absolute quantitative value has little meaning.

Thus, number of articles published, number of citations received, and *h*-index, totaled across all faculty members of an institution, determine BV for that institution. What about research funding, National Academy membership, and Nobel Prizes? My guess is these resources and awards flow to individuals and institutions with high BV. NA, NC, and HI represent the real accomplishments, as well as recognition of these accomplishments, that drive everything else at a research university, particularly for graduate programs. Nevertheless, my argument is that projections of BV are only proxies for rankings, not the actual rankings that, of course, include many other factors.

More on Metrics

Note that I have not included journal impact factors in this model. A journal's impact factor is the average number of citations per article published in the journal during the preceding 2 years. I certainly think that some journals are better than others, but impact factor emphasizes the wrong thing. A paper that receives an enormous number of citations but is published in a journal with low impact factor is a home run, not to be dismissed.

Seglen (1997) discusses why the impact factor of journals should not be used for evaluating research. He argues that "Use of journal impact factor conceals the difference in article citation rates—articles in the most cited half of articles are cited ten times as often as the least cited half." He also notes that "Journals' impact factors are determined by technicalities unrelated to the scientific quality of their articles. Further, journal impact factors depend on the research field: high impact factors are likely in journals covering large areas of basic research with a rapidly expanding but short lived literature that uses many references per article." Finally, and most important, he concludes, "Article citation rates determine the journal impact factor, not vice versa."

Van Noorden (2010) asserts that journal impact factor measures the popularity of the journal, not the quality of an individual contribution. He argues that the h-index is more relevant for assessing individual performance, but suffers from never decreasing, despite someone's productivity having plummeted. He discusses more elaborate metrics that assess highly cited papers that have been cited by highly cited papers, as well as measures that assess the centrality of a particular paper to a network of papers. He notes other metrics associated with number of online accesses.

Fitzsimmons and Skevington (2010) are concerned that journals with low impact factor are dismissed. They discuss how the calculation of impact factor is misleading, for example, "notes" are counted as articles, which inflates the denominator.

Nature (2010) considers the whole nature of bibliographic assessment. They conclude that the number of articles, number of citations, and h-index are objective and, therefore, can counter subjective biases. However, they should not be used exclusively. They note that impact factor is intended to evaluate journals, not individual contributors.

Braun (2010) reports on a panel of observations on how to improve the use of metrics. The consensus of the panel is that NA, NC, and HI should inform decision making about promotion and tenure (P&T), not determine it. Broader evaluations of contributions are needed. They suggest that P&T committees should actually read candidate's papers! The panel's bottom line—these metrics make sense but are too narrow.

Example

In Chapter 9, we compared the productivity of a hypothetical faculty member who only submitted articles to *Nature* and proposals to NIH and another that only submitted articles to *IEEE Transactions* and proposals to NIH. Due to the 4.5% annual increase of proposals submitted to NIH, both faculty members were required to write an exponentially increasing number of proposals to maintain constant funding. Thus their time spent preparing and submitting articles was steadily decreasing such that the faculty member focused on *Nature* published only 5 articles over 20 years, while the faculty members focused on *IEEE Transactions* published 15 articles during that period.

Now consider a university that has 50% tenure-track faculty members, in this example amounting to 183 faculty members. We will assume no growth of the student body and, hence, no growth in the number of faculty. Thus, BV will only increase due to publications of these 183 faculty members, the subsequent citations, and the evolving HI. Figure 10.2 shows the growth of BV, comparing the *Nature* and *IEEE Transactions* strategies.

The discontinuities of these curves are due to the discontinuous nature of HI. For the index to increase from N to $N+1$, it may take several years for the citations to accumulate.

Notice that both curves exhibit diminishing returns for two reasons. First, preparing NIH proposals is taking more and more time away from article writing. Second, the acceptance rates of both journals are steadily declining, due to annual increases of demand: 2.25% for *Nature* and 9% for *IEEE Transactions*.

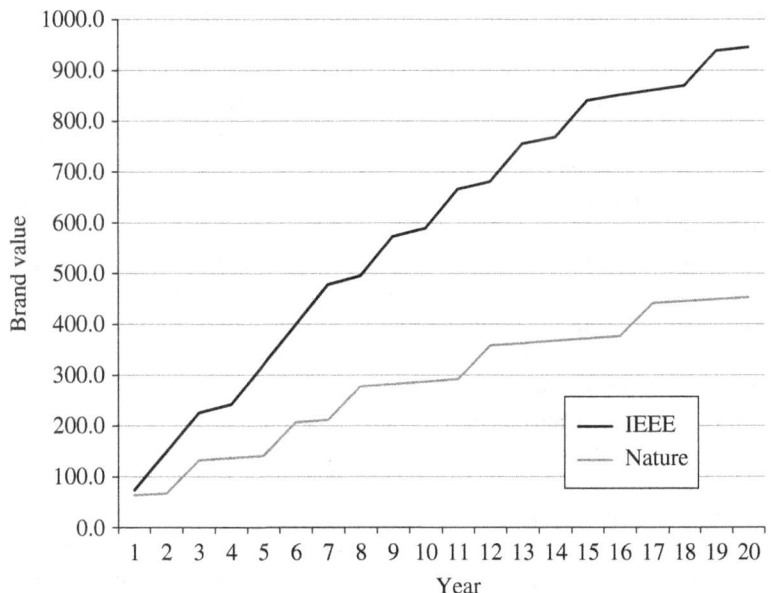

FIGURE 10.2 Growth of brand value.

IEEE Transactions led to greater BV because more papers are accepted and published, resulting in more citations and increasing HI. *Nature* is a prestigious publication but a faculty member cannot get enough articles published—especially when proposals require more and more time—to gain P&T. Further, as illustrated in Figure 10.2, this productivity paradox undermines the BV of the institution.

If we allowed the student body to grow in this example, then more classes would be offered and more faculty members would be hired, a portion of them tenure track. This could result in BV steadily growing rather than exhibiting diminishing returns. The BV per faculty member would have diminishing returns, but the overall growth would mask this phenomena. The likely result would be that the leadership of the university would not sense and address the productivity paradox.

As a consequence, the leadership would continue to advocate a traditional strategy, pursuit of which will progressively undermine the university's goals. I have found that all too often in academia faculty members have beliefs that are not supported by evidence. They do not exhibit this tendency in their own discipline, but they do in their beliefs about operation of the university as a whole. A key tenet of this book is the value of evidence-based decision making.

CONCLUSIONS

This chapter has addressed several ranking schemes that emphasize different attributes of a university. Georgia Tech was used as an example to illustrate the types of initiatives needed to improve rankings. Statistical analyses of rankings of one program—industrial/manufacturing/systems engineering—were reported. The best predictors of a program's ranking are last year's ranking plus the number of faculty.

The BV of a university was discussed as a proxy for rankings. An index of BV was presented that is a weighted sum of number of articles published, number of citations received, and h-index, totaled across all faculty members of an institution. An example was used to illustrate the nuances of the BV index as affected by research sponsors and publication outlets chosen by faculty members.

It is clear that research and education are the keys to world-class status for a university and economic growth for its key stakeholders. World-class status provides a wealth of opportunities for the university to serve the public. Economic development has long been a central element of such service. Universities' international initiatives are also becoming increasingly important.

Achieving world-class status requires excellent faculty and students, innovative programs, enormous resources, and lots of friends. The heterogeneity of research universities' constituencies makes it difficult to balance and satisfy competing interests. Consequently, Bok (2003) indicates that, even at Harvard, "There is never enough money to satisfy their desires" (p. 9).

This argues for the importance of vision and leadership in building a great university. World-class status is now much less likely to slowly emerge from intelligent people independently doing good work over many decades. This status is more likely to be attained when university leaders articulate focused strategies and cultivate the resources to attract the best people to pursue these initiatives.

REFERENCES

Alexander, J.H., Koenig, H.F., & Schouten, J.W. (2006). Building relationships of brand community in higher education: A strategic framework for university advancement. *International Journal of Educational Advancement, 6* (2), 107–118.

Bok, D. (2003). *Universities in the Marketplace: The Commercialization of Higher Education.* Princeton, NJ: Princeton University Press.

Braun, T. (2010). How to improve the use of metrics. *Nature, 465,* 870–871.

Clarke, M. (2002). Some guidelines for academic quality ratings. *Higher Education in Europe, 27* (4), 443–459.

Fitzsimmons, J.M., & Skevington, J.H. (2010). Metrics: Don't dismiss journals with low impact factor. *Nature, 466,* 179.

Gladwell, M. (2000). *The Tipping Point: How Little Things Can Make a Big Difference.* Boston, MA: Little Brown.

Gladwell, M. (2011). The order of things. *New Yorker, 87,* 68–75.

Hoyt, J.E., & Brown, A.B. (2003). Identifying college choice factors to successfully market your institution. *C&U Journal,* Spring, 3–10.

Lombardi, J.V., Phillips, E.D., Abbey, C.W., & Craig, D.D. (2012). *The Top America Research Universities: 2012 Report.* Phoenix, AZ: Center for Measuring University Performance, Arizona State University.

Marginson, S. (2007). Global university rankings: Where to from here? *Proceedings of the Asia-Pacific Association for International Education, National University of Singapore,* March 7–9, Singapore.

National Research Council. (1995). *Research-Doctorate Programs in the United States: Continuity and Change.* Washington, DC: National Academy of Sciences.

Nature (2010). Assessing assessment. *Nature, 465,* 845.

Noel-Levitz (2012). *Why Did They Enroll? The Factors Influencing College Choice.* Iowa City, IA: Ruffalo Noel Levitz.

Quiggin, J. (2015). Rank delusions. *The Chronicle of Higher Education,* February 16.

Rouse, W.B., & Garcia, D. (2004). Moving up in the rankings: Creating and sustaining a world-class research university. *Information Knowledge Systems Management, 4,* 139–147.

Ruhvargers, A. (2012). *Global University Rankings and Their Impact.* Brussels, Belgium: European University Association.

Seglen, P.O. (1997). Why the impact factor of journals should not be used for evaluating research. *British Medical Journal, 314,* 498–502.

Svoboda, P., & Harantova, M. (2014). Perceived quality of the higher education institution in relation to building strong brand from the alumni perspective. *Proceedings of the 14th International Academic Conference,* October 14, Malta.

US News & World Report [USN&WR]. (2003). Best Graduate Schools. Corporate Website: www.usnews.com (February 9, 2016).

Van Noorden, R. (2010). A profusion of measures. *Nature, 465,* 864–866.

11

TRANSFORMATION SCENARIOS

Higher education does not need to be tweaked here and there. A business process improvement initiative would help a bit, but be far from adequate. Instead, higher education needs to be fundamentally transformed to provide much greater value to many more people at much lower cost. Higher education needs to be transformed to address several alternative futures quite different from today. These futures are already emerging.

In this chapter, forces for change are discussed in terms of costs and benefits, globalization, and technology. Organizational change in higher education is then addressed, including concepts and principle drawn from domains other than higher education. Four alternative scenarios for the future of higher education are next elaborated. These scenarios provide the basis for considering transformation of academia. The chapter concludes with a discussion of historical perspectives on how change happens. This sets the stage for exploring the future in Chapter 12.

FORCES FOR CHANGE

Change happens when it is forced. The force can be an opportunity or a threat, perhaps embodied in a crisis. In many domains, the forces for change are manifested as market forces. Competitors, large or small, recognized or unrecognized, are the sources of market forces. Thus, change happens when there is competition to meet market needs with better performance, quality, service, and/or prices.

Universities as Complex Enterprises: How Academia Works, Why It Works These Ways, and Where the University Enterprise Is Headed, First Edition. William B. Rouse.
© 2016 John Wiley & Sons, Inc. Published 2016 by John Wiley & Sons, Inc.

Change is much less likely to happen when there is little or no competition. The stewards of the status quo can then argue that everything is great—as good as it can be within the prevailing constraints. If competitors are not offering viable alternatives, it is difficult to convincingly argue that the stewards of the status quo are wrong. Consequently, all investments are focused on preserving the status quo.

As evidenced by many discussions in earlier chapters, higher education faces an emerging crisis. Costs of administration (per student) have soared, while expenditures on education and research (per student) have slowly declined. At the same time, the merits of the old argument that a college education is always worth it have faded. Unfortunately, there has been a dearth of viable alternatives. Online education has some benefits, but campus-based education provides a much richer growth experience. However, the technology will inevitably get better and better.

Costs and Benefits

The Economist (2014a) foresees creative destruction in higher education precipitated by a cost crisis, changing labor markets, and new technology. The result, they argue, will turn an old institution on its head. They note that "In America, government funding per student fell by 27% between 2007 and 2012, while average tuition fees, adjusted for inflation, rose by 20%." Forty-seven percent of students in America do not finish their degree programs. For many majors, a college education no longer results in a good job. Technology is the third driving force, as discussed in the succeeding text.

The Economist (2014b) also addresses the future of universities. They discuss the funding crisis in terms of "cost disease" that was discussed in Chapter 6. The softening of jobs for graduates has been exacerbated by technology that has automated many jobs once performed by humans.

The technological revolution is also affecting the delivery of education. As an example, they cite Georgia Tech's $7000 online MS degree in computer science, compared to the exact same degree program on campus for $25,000. They project that the massive open online course (MOOC) offerings will inevitably succeed. In a related article (Economist, 2014c), they review Brazil's private online universities that have three-quarters of the market with low fees and rapidly increasing quality.

These three articles together assess the costs and benefits in higher education to be going in the wrong direction. They quote Clayton Christensen's comment that within 15 years half of traditional universities will be in bankruptcy. Clearly, big changes, for better or worse, will happen. In this chapter, I argue that transformation of the higher education enterprise can be approached systematically, increasing the chances of better outcomes.

Globalization

Wildavsky (2010) discusses how international competition for the brightest minds is transforming the world of higher education. Newly created or expanded universities in Asia are competing with Harvard, MIT, Cambridge, and Oxford for faculty, students, and research preeminence. Satellite campuses of Western universities are emerging all over the world. He chronicles the unprecedented international mobility of students and faculty, the spread of branch campuses, and the international expansion of college rankings. He argues that this scholarly marketplace will spread knowledge benefits to everyone, both educationally and economically.

The Economist (2011) highlights foreign students getting advanced degrees in America. They profile MIT in particular and its connections with global industries. American-educated faculty members are increasingly joining faculties of universities around the world. This echoes the projections of Wildavsky.

Altbach and Salmi (2011) report on a World Bank-funded case-based study of the evolution of 11 research universities in 10 countries—2 in Chile and 1 each in China, Hong Kong, India, Korea, Malaysia, Mexico, Nigeria, Russia, and Singapore. In an introductory chapter, Altbach (2011) focuses on research universities and, in particular, the 150 globally relevant research universities of 4800 postsecondary institutions in the United States. He notes that "Both faculty members and students are increasingly recruited internationally, and mobility is now an established fact of contemporary higher education, especially affecting research universities." He also asserts that research universities "must recognize the primacy of merit, and their decisions are based on a relentless pursuit of excellence." Consequently, perhaps 90% of the articles published in top journals come from faculty members of research-intensive universities.

Altbach considers several current and future challenges. As always, funding is an issue. Autonomy is an increasing challenge as externally imposed accountability and compliance are encroaching. Attracting the best and brightest faculty members and graduate students requires competing with other sectors of the economy that provide better salaries. Globalization has increased the fierceness of the competition for talent. Increased privatization, in terms of using the marketplace as a source of funds, risks the possibility of external factors driving the agenda.

In a concluding chapter, Salmi (2011) begins by noting that the top 10 research universities in the world were all formed before 1900 and 2 are more than eight centuries old. He then elaborates common themes from the 11 much newer research universities studied. Attracting, recruiting, and retaining talented academics were challenges for all of them. Needs to be well resourced was a common theme. Governance issues, particularly protection from political interference, were mutual concerns. Considering paths of development, common elements included:

- Relying on the diaspora to recruit native born, but foreign-educated faculty members to return
- Employing English as the main language for both education and research
- Concentrating on science and engineering rather than all disciplines
- Using benchmarking to inform improvement programs
- Adopting curricula and pedagogical innovations

The Royal Society (2011) provides an assessment of global scientific collaboration in the twenty-first century. They begin by noting that there are 7 million researchers worldwide supported by $1 trillion, a 45% increase since 2002.

They report that science is increasingly global. Science in China, India, and Brazil is increasing faster than the global trend, but the United States, Europe, and Japan still lead the pack. They note that scientific activity is concentrated in a number of widely dispersed hubs. Beyond these hubs, however, scientific activity is also flourishing where economic impact is recognized and understood.

They explain that the scientific world is becoming increasingly interconnected with growing international collaboration. Collaboration enhances quality, improves efficiency and effectiveness, and is increasingly often necessary. The primary drivers of collaboration are scientists themselves. Collaborative networks are motivated by bottom-up exchange of insights, knowledge, and skills, leading to science becoming global rather than national. Collaboration brings significant benefits, for instance, increased citations of participants' publications.

Recognized global challenges are now key components of national and multinational strategies and funding. Global challenges are interdependent and related to each other. A variety of organizational models have been used to address these challenges, and, they assert, there are many lessons to be learned. They argue that science is essential for addressing global challenges but cannot do this in isolation. They conclude that all countries have a role in pursuit of these challenges.

Bavorick and colleagues (2011), reporting on the Royal Society study, emphasize the observation that China may overtake the United States in scientific output by 2013. The implications of this trend are not clear, as the United States retains a clear lead in translating research results into economic activities and returns.

The National Academies (2007) argue that the growth of US economy is primarily driven by technological innovation. They assert that globalization is threatening the US leadership position. Leading-edge science and engineering is being accomplished in many places around the world. Ten recommendations are provided to enhance the US science and technology enterprise. Not surprisingly, requests for government funding are laced through these recommendations. Earlier examples of such lines of reasoning include the National Academies (1995) discussion of the impact of the end of the Cold War and decreased funding of university research by DoD. This echoes a stream of National Research Council reports (NRC, 1985, 1987, 2012) that argue the benefits of research and the need for more federal funding.

A Tsunami of Talent

I chaired the International Review Board at Tsinghua University in Beijing, China, where we conducted a review of the Department of Industrial Engineering. They presented a few basic statistics that were of particular note.

In China, 40% of undergraduates matriculate in engineering. At Tsinghua at least, of those that graduate, 67% continue to graduate school. While it is risky to generalize, this leads to an estimate that 25% of college graduates in China have advanced degrees in engineering.

In contrast, 4% of US undergraduates matriculate in engineering. Of those that graduate, 12% continue to graduate school. Thus, roughly 0.5% of college graduates in the United States have advanced degrees in engineering.

I may be off, perhaps by quite a bit, but 0.5 versus 25% is a rather daunting difference. If this compounds year after year, decade after decade, the consequences could be astounding.

Technology

DeMillo (2011, 2015) discusses a panorama of technology possibilities. Computer-based instruction has a long history, perhaps beginning with PLATO at the University of Illinois in the 1960s. Control Data Corporation commercialized PLATO in the 1970s. I experienced it as a faculty member at Illinois in the mid-1970s. I remember it as very impressive but also very expensive.

Earlier, I discussed our forays into simulation-oriented computer-based instruction for operators and maintainers of complex systems. A key enabler was inexpensive computing power, namely, the Apple II and later the IBM PC. Via this technology, our professional training offerings became inexpensive and portable.

The Internet, and more recently portable smart devices, has taken the possibilities to new levels. DeMillo argues that "Higher education is an Internet business." A professor can now teach hundreds of thousands or even millions of students online with offerings ranging from Stanford University's computer science courses to Salman Khan's short videos teaching basic math concepts.

Ho and his colleagues (2015) report on 2 years of experiences of HarvardX and MITx. They indicate that growth is steady in overall and multiple-course participation in HarvardX and MITx. They report that over 2 months, there were 1300 unique participants per day yielding 1.03 million in total, accounting for 1.71 course entries per participant.

They found that participation initially declines in repeated courses but then stabilizes. Surveys suggest that a slight majority intends to apply for certification. Many of these are teachers. Participation and certification differ by curricular area; for example, participation in computer science courses is four times higher than other areas. Course networks reveal the centrality of large computer science courses and the potential of sequences of modules. Certification rates are high among those who pay $25–$250 to "ID-verify" their certificates.

Nevertheless, there are skeptics. Toyama (2015) argues that technology will never fix education. He asserts that "Even when technology tested well in experiments, the attempt to scale up its impact was limited by the availability of strong leadership, good teachers, and involved parents." Further, "Any positive effects depend on well-intentioned, capable people." Finally, "The real obstacle in education remains student motivation."

ORGANIZATIONAL CHANGE

Enterprise transformation involves substantial organizational change rather than just business process improvement (Rouse, 2006). This section first examines a few recent treatises on needed changes in academia. I then discuss theories and practices of organizational change more broadly.

Christensen and his colleagues (2011) present the usual summary of increasing costs of higher education and diminishing performance. They assert that we need to shift focus from making traditional college accessible to more students to making quality postsecondary education affordable. They argue for disruptive innovation via technology and new business models. These new models, they assert, will have to be imposed on universities because they will not be able to change on their own. This is due to their currently having three fundamentally incompatible business models for education, research, and service. They conclude that elite research institutions should continue what they are doing; everybody else should focus on affordable education. This includes needs to "escape from the policies that focus on credit hours and seat time to one that ties progression to competency and mastery."

Taylor (2010) presents a plan for reforming colleges and universities. His suggestions include ending tenure, restructuring departments to encourage greater cooperation, emphasizing teaching rather than research, and bringing education to new students, using online networks to connect these students worldwide. The resulting efficiencies and innovations, he argues, will help graduate students deal with an unpromising job market and undergraduates with massive burdens of debt.

Khurana (2010) argues that university-based business schools were founded to train a professional class of managers similarly to the training of doctors and lawyers. He asserts that business schools have retreated from this goal. Transforming business into a profession required codifying the knowledge relevant for practitioners and developing professional norms. He traces how business educators confronted these challenges. He argues that business schools have largely stopped being concerned about professionalism and have become merely purveyors of a product, the MBA, with students treated as consumers. A perspective that managers' sole responsibility is shareholder value has replaced professional and moral ideals that once animated and inspired business schools.

Menand (2010) addresses reform and resistance in the American university. He characterizes the core of resistance by asserting that "Since it is the system that ratifies the product—ipso facto, no one outside the community of experts is qualified to rate the value of the work produced within it—the most important function of the system is not the production of knowledge. It is the reproduction of the system. To put it another way, the most important function of the system, both for purposes of its continued survival and for purposes of controlling the market for its products, is the production of the producers. The academic disciplines effectively monopolize (or attempt to monopolize) the production of knowledge in their fields, and they monopolize the production of knowledge producers as well." He concludes that "The key to reform of almost any kind in higher education lies not in the way that knowledge is produced. It lies in the way that the producers of knowledge are produced."

Theory and Practice

Hannan and Freeman (1984) address structural inertia and organizational change. They note that "Selection pressures in modern societies favor organizations whose structures are resistant to change." They discuss several phenomena common to organizational change:

- Structural inertia increases with age and size, while organizational death rates decrease with age and size
- Attempting reorganization decreases reliability of performance, while it increases organizational death rates
- Structural reorganization produces a liability of newness, and the duration of the reorganization increases death rates of organizations
- Complexity increases the expected duration of reorganization and increases the risk of death due to reorganization

Reviewing this list, it is easy to see why universities have immense difficulties addressing needs to change, and how they approach change in very drawn-out manners, and consequently face great risks of complete failure.

Sternberg (2012) considers why colleges may not be able to change. He argues that five elements are required for change:

- Ability to change—by the time the need to change is recognized, the organization may no longer have the needed resources
- Belief in the ability of the organization to change—this requires avoiding the belief of being stuck
- Desire to change—this requires being open to changing core beliefs and practices
- Desire to appear to change—this requires being open to changing an assumed image, which may have faded long ago
- Courage to translate ideas into action—dealing with vested interests that do not want change

I have encountered these elements many times when working with universities, companies, and government agencies. These experiences led me to formulate the needs–beliefs–perceptions model (Rouse, 1993). In the context of this book, this model suggests that faculty members and administrators have various beliefs they need to be true. For example, they believe that NIH and NSF funding is most impressive because they need their career paths and choices to be validated. To the extent that their beliefs are based on data, they will have sought data that support these beliefs. For the most part, however, needs and beliefs are completely experiential rather than evidence based. This can really hinder change.

I have worked with well over 100 universities, companies, and agencies, most of them large and well known. I used to keep track of the executives and senior managers with whom I worked. I stopped doing this when the database got to 5000 people. Reflecting on these experiences led me to identify a set of common organizational delusions (Rouse, 1998). For universities, delusions take the form of faculty members and administrators persisting with assumptions that are far from warranted. These assumptions typically are associated with trends in the external world, what competitors are doing and intend to do, internal competencies and abilities to execute, and alignment of incentives and rewards with goals and strategies. Delusions in these areas can completely undermine any organizational change initiatives.

I returned to academia after 13 years of founding and growing two companies. My research agenda had shifted to enterprise transformation. In particular, I focused on understanding and enabling fundamental change. This led to formation of the Tennenbaum Institute and compiling theories, practices, and case studies of transformation (Rouse, 2006). The resulting transformation framework is introduced after consideration of four scenarios for the future of academia.

FOUR SCENARIOS

This section elaborates four scenarios for the future, which I developed for strategic planning engagements with several universities. Clash of Titans is basically "business as usual on steroids." The top-ranked universities continue to compete for elite status, the best faculty and students, and lucrative grants and contracts. Hot, Flat, and Crowded borrows Tom Friedman's phrase and addresses global universities competing very successfully with top programs in the United States and Europe. This has enormous financial implications for American and European programs. Lifespan Mecca reflects the steadily increasing average age of graduate students. Such students will not accept the service levels endured by 18–22-year-olds. For example, they will not put up with large classes and will demand access to faculty members. Network U. addresses the pervasive impact of technology on education, leading to everyone taking physics, chemistry, economics, and so on from the best faculty anywhere. Employment of teaching faculty will plummet.

These four scenarios are elaborated in this section in considerable detail. The abilities of universities to respond to these alternative futures are discussed in terms of the material presented in the earlier chapters. This discussion concludes with careful consideration of the prospects for change of the academic enterprise. These alternative futures are computationally explored in Chapter 12.

What will the academic world be like in 20 years—2035? Thinking 20 years into the future is quite difficult, as is evidenced by thinking back to 1995 and imagining our current iPhones, Kindles, and pervasive social technology such as Facebook.

Nevertheless, it is interesting—and potentially useful—to consider future scenarios. We know one thing for sure. Any one scenario will inevitably be wrong. Thus, we need multiple scenarios.

Driving Forces

Scenario development should be based on best practices on this topic (Fahey & Randall, 1998; Schoemaker, 1995; Schwartz, 1991). All of the pundits begin by defining the forces that drive the future. There are—at least—four strong driving forces that will affect academia's future:

1. Competition among top universities will become increasingly intense, both for talent and resources—there will be a clash of the titans.
2. Globalization will result in many academic institutions, particularly in Asia, achieving parity in the competition—it will become hot, flat, and crowded.
3. Demographic trends portend an aging but active populace leading to an older student population—higher education will need to become a lifespan mecca.
4. The generation of digital natives will come of age, go to college, and enter the workforce—there will be no choice but become a networked university.

We cannot escape these forces, nor can we fully predict the ways in which they will interact to shape the world of 2035. We can be sure, however, that for academic institutions to compete in this future, their strategies must be sufficiently robust to accommodate these forces. If, instead, they focus on just one scenario—for example, the Clash of Titans that most closely resembles business as usual, perhaps on steroids—they will almost certainly be at a competitive disadvantage in the future.

Clash of Titans

I have worked at, consulted with, or served on advisory boards of quite a few top universities. Every one of them pays attention to their *US News & World Report* ranking. They aspire to battle with the titans of higher education and hold their own. This scenario has universities continuing that clash, perhaps clawing their way to higher rankings, albeit in an increasingly competitive environment.

General Description: Academic institutions continue to battle to achieve dominance in various academic disciplines, as well as competing with top universities for overall rankings within the United States and with premier international universities for global rankings.

Dominant Issues: The competition for talent becomes fierce, with well-endowed chairs becoming the minimum for attracting faculty talent; top students at all levels expect and get near-free education.

Economic Implications: The top players continue to dominate receipt of federal funds, with considerable pushback from other players; costs of facilities and labs soar, much of which must be raised from philanthropic sources.

Social Implications: University cultures are sustained, with adaptations for a decreasingly Caucasian male population—for both students and faculty—but one that is committed to the values and sense of purpose that has been central for recent decades; changing demographics impacts how alumni best relate to their alma maters.

Hot, Flat, and Crowded

Tom Friedman (2005) has argued that the world is flat and we should no longer assume business as usual—his revision of this best seller included a chapter on Georgia Tech and how they transformed education in computing. More recently, Friedman (2008) has argued that the world will be hot, flat, and crowded. In this scenario, academic institutions have to compete with a much wider range of players in a global arena.

General Description: Global parity emerges in graduate education in science and technology, particularly for traditional disciplines and subdisciplines; greater collaboration among institutions emerges; demand for higher education in the United States will nevertheless increase substantially.

Dominant Issues: Many of the best jobs are in Asia; scarcity and constraints dominate sustainability debates; clashes of belief systems create political turmoil and security concerns; meeting demands presents strong challenges.

Economic Implications: Federal and state support diminish as portions of budget; industrial and philanthropic support is increasingly competitive; sponsors become sensitive to where resources are deployed; undergraduate tuition stabilizes and increases are less and less acceptable.

Social Implications: Global footprints of top universities increase by necessity; social, cultural, and ethnic diversity of faculty and students increases in turn; traditional business practices, for example, promotion and tenure, must change to accommodate diversity.

Lifespan Mecca

It is easy—and convenient—to assume that the students of the future will be much like the students of today. However, CSGNET (2007) reports that over the past decade the number of graduate students 40 years old and older has reached record numbers. From 1995 to 2005, the number of postbaccalaureate students aged 40 and older at US colleges and universities jumped 27%. And during the next two decades, the number of older citizens will rise at even faster rates than the number of those 24 and younger, which suggests that the number of postbaccalaureate students aged 40 and over very likely will continue to grow. In this scenario, universities have to address a "student" population with more diverse interests and expectations rather different from students of the past and current eras.

General Description: Demand for postgraduate and professional education surges as career changes become quite common; demand steadily grows for education and artistic performances by an increasingly urban older population.

Dominant Issues: Two or three MS or MA degrees become common across careers, as do often required certificate programs; multiple artistic performance and sporting events per day become common at any top university.

Economic Implications: Tuition revenues soar for professional programs and graduate education programs popular with elders; revenues from artistic performance and sports venues become significant portions of university budgets.

Social Implications: The median age of students increases substantially, changing the campus culture markedly; older students in particular expect and get high-quality, user-friendly services; diversity of faculty increases significantly to satisfy diversity of demands.

Network U.

Technology is increasingly enabling access to world-class content in terms of publications, lectures, and performances. Higher education can leverage this content to both increase quality and lower costs (Kamenetz, 2009; North, 2009). This technology has also spawned the generation of "digital natives" that is always connected, weaned on collaboration, and adept at multitasking. In this scenario, academia has to address different types of student using very different approaches to delivering education and conducting research.

General Description: Social technology prevails; access to the best content and faculty is universal; nevertheless, students go to college to learn and mature; however, the classroom experience is now highly interactive, both remotely and face to face.

Dominant Issues: Students and faculty have broad and easy access to knowledge, often via other people; with the "best in class" universally available, local faculty play more facilitative roles in small (10–20) "high touch" discussion groups.

Economic Implications: More teaching professionals are needed for recitation-sized classes; teaching skills are at a premium; increasing numbers of high-quality programs result in strong downward pressure on tuition and fees; faculty research becomes near totally externally funded.

Social Implications: Students and faculty are networkers *par excellence*, both within and across institutions; students' evaluations of teaching effectiveness play an increasing role; students seamlessly transition from K–12 to university to lifespan education.

Implications

Framing the future 20 years from now is quite difficult. Yet, this is essential if academic institutions are to focus their competencies and resources on the possible futures in which our students—and all of us—will have to compete. Our abilities to understand and manage the inherent uncertainties associated with these futures can be an enormous competitive advantage. We need to enhance these abilities to maintain our competitive position in global education.

TRANSFORMING ACADEMIA

There is a wide range of ways to pursue transformation. Figure 11.1 summarizes conclusions drawn from numerous case studies (Rouse, 2006). The ends of transformation can range from greater cost efficiencies to enhanced market perceptions, to new product and service offerings, and to fundamental changes of markets. The means can range from upgrading people's skills to redesigning business practices, to significant infusions of technology, and to fundamental changes of strategy. The scope of transformation can range from work activities to business functions, to overall organizations, and to the enterprise as a whole.

The framework in Figure 11.1 has provided a useful categorization of a broad range of case studies of enterprise transformation. Considering transformation of markets, Amazon leveraged IT to redefine book buying, while Wal-Mart leveraged IT to redefine the retail industry. In these two instances at least, it can be argued that Amazon and Wal-Mart just grew; they did not transform. Nevertheless, their markets were transformed.

Illustrations of transformation of offerings include UPS moving from being a package delivery company to a global supply chain management provider, IBM's transition from manufacturing to services, Motorola moving from battery eliminators to radios to cell phones, and CNN redefining news delivery. Examples of transformation of perceptions include Dell repositioning computer buying, Starbucks repositioning coffee purchases, and Victoria's Secret repositioning lingerie buying.

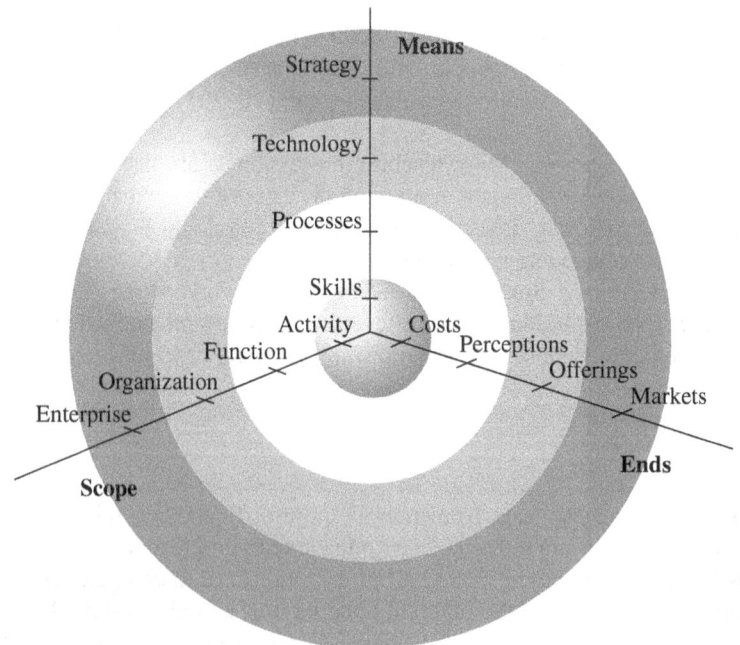

FIGURE 11.1 Transformation framework.

The many instances of transforming business operations include Lockheed Martin merging three aircraft companies, Newell Rubbermaid resuscitating numerous home product companies, and Interface adopting green business practices.

The costs and risks of transformation increase as the endeavor moves farther from the center in Figure 11.1. Initiatives focused on the center will typically involve well-known and mature methods and tools from industrial engineering and operations management. In contrast, initiatives toward the perimeter will often require substantial changes of products, services, channels, etc., as well as associated large investments.

It is important to note that successful transformations in the outer band of Figure 11.1 are likely to require significant investments in the inner bands also. In general, any level of transformation requires consideration of all subordinate levels. Thus, for example, successfully changing the market's perceptions of an enterprise's offerings is likely to also require enhanced operational excellence to underpin the new image being sought. As another illustration, significant changes of strategies often require new processes for decision making, for example, for R&D investments.

The transformation framework can be applied to thinking through the four scenarios for academia. At this point, I will only address what is likely to have to change, not how the changes can be accomplished. In particular, I will not consider how to gain the support of stakeholders, manage their perceptions and expectations, and sustain fundamental change (Rouse, 2001, 2006, 2007).

Clash of Titans

This scenario involves changing perceptions across the research function and associated organizations to continually enhance brand value and, hence, move up in the rankings. This will inevitably increase costs to subsidize research. Increased government funding will be difficult to secure, so emphasis will shift to industry funding as well as gifts and endowment. Increased industry funding will require expanding offerings and enabling processes to include services to sponsors, perhaps including professional education. This will, of course, affect the education function and associated organizations.

Of the four scenarios, this one is the least disruptive because the focus is on enhancing the enterprise's abilities to do what it is already doing. Processes and skills will be the primary means. It may be possible to leave much of the enterprise unchanged, although the financial resources needed by the research function to make it up in the rankings may have to come, in part, from cost reductions in other parts of the enterprise.

Hot, Flat, and Crowded

Universities will need to decrease dependence on foreign students, particularly in graduate programs. Unfortunately, many of these graduate programs tend to be among the most profitable of a university's offerings. The needed changes will be in the education function and related organizations. The university will need to attract more top US students to graduate school, although most of them are not inclined toward graduate school, typically because the financial sacrifice is too great. Another strategy is expanded professional education offerings to attract employees of US firms with tuition reimbursement programs. This will require increased nontenure-track faculty members who have the industry experience needed to deliver professional education. Another alternative is to decrease graduate offerings, but this will require substantial decreases of costs.

The means to accomplish these ends may only require enhanced processes and skills. However, competing in the global higher education marketplace may also require a significant change of strategy and enabling technologies. For example, delivery of programs might involve a mix of on-campus classes and distance education, perhaps blended with case studies in other countries to give the students, for example, experiences of global logistics hubs or alternative healthcare delivery processes. This would give a university's offerings a decidedly global flavor.

Lifespan Mecca

This scenario will require increased MS and MA offerings, probably not taught by tenure-track faculty. Older students will expect smaller classes with increased opportunities for interaction. They will also be looking for enhanced learning experiences with less emphasis on grades and credentials. Universities will have to change perceptions of the education function to reflect these changes. Costs will have to be decreased significantly, as tuition will seldom be reimbursed.

These changes will affect the education function and related organizations. The primary means will be enhanced processes and skills. The success of these new offerings may depend on changing perceptions of the university's mission and vision. This may involve a bit of rebranding to convince older students that the university understands their needs and wants to meet them.

Network U.

This scenario will require the most pervasive changes. It basically involves changing offerings across the instruction function and associated organizations via process and technology changes. Put simply, teaching will be quite different.

Richard Feynman would teach all students physics. Paul Samuelson would teach them economics. The fact that both of these individuals are deceased will not matter given the state of digital media technology. They would even be able to lecture on research results that were not known before they died.

Students would no longer sit in crowded auditoriums, while professors lectured on the stage. The lectures could be viewed when and where students chose. They could watch and listen to the lectures again and again. Small groups could gather around the presentation media and occasionally put the lecture on hold and discuss the points just made.

The vast majority of teachers would no longer prepare lecture notes and deliver lectures. Instead, they would meet with small groups of students, say, 10–20, to discuss the lectures and talk about the implications of the material. Most teachers would become recitation leaders rather than lecturers. They would be "guides on the side" rather than "sages on the stage."

This would also change the bricks-and-mortar strategies of universities. Few large lecture rooms would be needed, but many more small meeting rooms would be required. Perhaps faculty member offices would become more like elementary school classrooms with a work area to one side and, in the middle, seating for the 10–20 students in each recitation group. A faculty member might meet with four to six of these groups each day.

There would be few teaching assistants because the faculty members would now be responsible for far fewer students. The demands of correcting exams and homework would be eased by technology, including social technology that would engage students in this process. Collaborative learning would have become the norm for the generations of "digital natives."

In parallel, full-time researchers will pursue focused programs of research. The younger research faculty members will not have tenure. Research will become more networked and collaborative, both across disciplines and across academia, industry, and government. Funding from industry will increase substantially, while government funding will decline due to other demands on these resources. Networked collaboration technology will be a great enabler, both for work and for translation of research outcomes to practice via the industry sponsors.

HOW CHANGE HAPPENS

In thinking about how changes such as outlined earlier might actually happen, it is useful to consider how change differs within various aspects of society. Are differing changes somehow related? C.P. Snow has argued that there is a chasm between the arts and humanities and science and technology (Snow, 1965). However, all of these endeavors are inevitably influenced by the times in which they are pursued.

Consider the late eighteenth and early nineteenth centuries. Richard Holmes (2008) describes the lives and scientific accomplishments of Joseph Banks, William Herschel, and Humphry Davy during the "second scientific revolution" of 1770–1830. He outlines how their popularization of findings in exploration, botany, astronomy, and chemistry influenced the poetry of their contemporaries: Coleridge, Byron, Keats, and Shelley.

Szollosi-Janze (2005) discusses transformation of research and education in the period between the Wilhelmine Empire and the Weimar Republic in Germany. He emphasizes the close cooperation between government, industry, and science during 1880–1920. It is rare that the academic enterprise can undertake fundamental change by itself.

Moving to the later nineteenth and early twentieth centuries, Louis Menand (2001) presents a study of Oliver Wendell Holmes, William James, Charles Sanders

Peirce, and John Dewey that shows how these four men developed a philosophy of pragmatism following the Civil War and continuing, at least, until World War I. Their thinking fundamentally affected America in law, science, and education, reaching far beyond academia.

Finally, considering the early and mid-twentieth century, Howard Gardner (1994) portrays the twentieth-century creative genius of Sigmund Freud, Albert Einstein, Pablo Picasso, Igor Stravinsky, T.S. Eliot, Martha Graham, and Gandhi. Arthur Miller (2002) focuses on just Einstein and Picasso. He chronicles the impact of Henri Poincaré's 1902 geometry book *La Science et L'Hypothèse* on the thinking of these two great geniuses. The non-Euclidian exposition in this book influenced both the notion that gravity bends light (i.e., relativity) and the Cubist movement in art. These are two seminal developments in the early twentieth century.

Thus, change in one arena can have enormous impacts on other areas, sometimes directly but often indirectly as such changes are manifested in the broader social dialogue. In this way, the causality of change often functions more like a network of relationships through which change propagates, rather than a simple *A* affects *B*.

We cannot know what mix of the four scenarios will play out in higher education over the next couple of decades. As DeMillo (2011, 2015) chronicles, many innovators are already experimenting with elements of these scenarios. It is also quite likely that elite institutions will stick with Clash of Titans, at least initially. However, as some mix of the other three scenarios plays out across the nonelite institutions, it will be difficult for the elite to avoid popular innovations in educational effectiveness and efficiency. My guess is that the next decade or two will later be characterized as a revolution in higher education, as well as more broadly in society.

CONCLUSIONS

Michael Lewis (2004) relates the story of the Oakland Athletics and their ability to use scientific management to maximize wins per dollar. The essence of the book's argument is that many of the truths that organizations embrace and use to guide decisions are, in fact, myths with no empirical basis in fact. Once you look at the data in detail, you can find what really matters. If your competitors continue to embrace the old (false) truths while you embrace the new empirically based truths, you can gain an enormous competitive advantage.

Of course, the priesthood associated with the old (false) truths will do their best to defend the dogma and discredit the new empirically based truths, often without even paying attention to the source and nature of the new truths. They will attack the integrity and abilities of those presenting the new truths, typically dismissing them as uninformed and self-serving.

I know that a university is much more complex than a baseball team, but I wonder if we are not often trapped by our assumed truths rather than empirically exploring what really matters and how the allocation of our resources could truly improve the value we provide.

Enterprise transformation should be evidence based. The evidence should inform what new things are initiated. Equally important, evidence should be used to decide what elements of the enterprise to keep and what elements to shed. Considering what to keep when transforming an academic enterprise, the following conclusions seem warranted:

- Keep the education line of business, but consider a much broader range of approaches to delivery; be cautious when investing in physical classrooms
- Keep the research line of business, but get the economics right to generate both knowledge and money; be skeptical of low probability opportunities
- Keep the service line of business that relates directly to the education and research businesses; spin off or outsource all the rest

Success in adopting this strategy will depend on several other things:

- Move to activity-based cost accounting and minimize nonattributable overhead costs; aspire to achieve a near-zero overhead rate
- Price services based on costs directly attributable to these services; include profit margins that are competitive in relevant markets
- Retain money-losing services only to the extent that they are vital to one of more lines of business; if there are many of these, you have not faced reality
- Outsource everything that someone else can perform better and/or cheaper; become expert at selecting and managing vendors and partners

There is one final critical need. Define, measure, and reward performance in all aspects of the business. This can be problematic in academia. Universities have great difficulty penalizing poor performance and even greater difficulty rewarding good performance. Thus, poor performers hang around—for years, even careers—and good performers get frustrated and leave. Fix this as soon as possible.

In this chapter, forces for change were discussed in terms of costs and benefits, globalization, and technology. Organizational change in higher education was then addressed, including concepts and principles drawn from domains other than higher education. Four alternative scenarios for the future of higher education were next elaborated. These scenarios provide the basis for considering transformation of academia. The chapter concluded with a discussion of historical perspectives on how change happens. This has set the stage for exploring the future in Chapter 12.

REFERENCES

Altbach, P.G. (2011). The past, present, and future of the research university. In P.G. Altbach & J. Salmi, Eds., *The Road to Excellence: The Making of World-Class Research Universities* (Chapter 1). Washington, DC: World Bank.

Altbach, P.G., & Salmi, J. (Eds.). (2011). *The Road to Excellence: The Making of World-Class Research Universities*. Washington, DC: World Bank.

Bavorick, H., Sanburn, J., Silver, A., & Tharoor, I. (2011). Another way China may beat the US. Time, February 14.

Christensen, C.M., Horn, M.B., Caldera, L., & Soares, L. (2011). *Disrupting College: How Disruptive Innovation Can Deliver Quality and Affordability to Postsecondary Education*. Washington, DC: Center for American Progress.

CSGNET (2007). *Data Sources: The Rise of "Older" Graduate Students*. http://www.cgsnet.org/portals/0/pdf/DataSources_2007_12.pdf (accessed August 1, 2015).

DeMillo, R.A. (2011). *Abelard to Apple: The Fate of American Colleges and Universities*. Cambridge, MA: MIT Press.

DeMillo, R.A. (2015). *Revolution in Higher Education: How a Small Band of Innovators Will Make College Accessible and Affordable*. Cambridge, MA: MIT Press.

Economist (2011). The global campus: The best universities now have worldwide reach. The Economist, January 20.

Economist (2014a). Creative destruction: A cost crisis, changing labor markets, and new technology will turn an old institution on its head. The Economist, June 28, 11.

Economist (2014b). The future of universities. The Economist, June 28, 20–22.

Economist (2014c). The higher education business: A winning recipe. The Economist, June 28, 59.

Fahey, L., & Randall, R.M. (Eds.). (1998). *Learning From the Future: Competitive Foresight Scenarios*. New York: Wiley.

Friedman, T.L. (2005). *The World Is Flat*. New York: Farrar, Straus and Giroux.

Friedman, T.L. (2008). *Hot, Flat, and Crowded*. New York: Farrar, Straus and Giroux.

Gardner, H.E. (1994). *Creating Minds: An Anatomy of Creativity as Seen Through the Lives of Freud, Einstein, Picasso, Stravinsky, Eliot, Graham, and Gandhi*. New York: Basic Books.

Hannan, M.T., & Freeman, J. (1984). Structural inertia and organizational change. *American Sociological Review, 49* (2), 149–164.

Ho, A.D., Chuang, I., Reich, J., Coleman, C., Whitehill, J., Northcutt, C., Williams, J.J., Hansen, J., Lopez, G., & Petersen, R. (2015). *HarvardX and MITx: Two Years of Open Online Courses*. Cambridge, MA: HarvardX, Working Paper No. 10.

Holmes, R. (2008). *The Age of Wonder*. New York: Vintage.

Kamenetz, A. (2009). Who needs harvard? Fast Company, September.

Khurana, R. (2010). *From Higher Aims to Hired Hands: The Social Transformation of American Business Schools and the Unfulfilled Promise of Management as a Profession*. Princeton, NJ: Princeton University Press.

Lewis, M. (2004). *Moneyball: The Art of Winning an Unfair Game*. New York: Norton.

Menand, L. (2001). *The Metaphysical Club: A Story of Ideas in America*. New York: Farrar, Straus and Giroux.

Menand, L. (2010). *The Marketplace of Ideas: Reform and Resistance in the American University*. New York: Norton.

Miller, A.I. (2002). *Einstein, Picasso: Space, Time, and the Beauty That Causes Havoc*. New York: Basic Books.

National Academies (1995). *Forces Shaping the U.S. Academic Engineering Research Enterprise*. Washington, DC: National Academy Press.

National Academies (2007). *Rising Above the Gathering Storm: Energizing and Employing America for a Brighter Economic Future*. Washington, DC: National Academy Press.

North, G. (2009). MIT Calls Academia's Bluff. www.LewRockwell.com (accessed February 9, 2016).

NRC (1985). *Engineering Graduate Education and Research: Engineering Education and Practice in the United States*. Washington, DC: National Academy Press.

NRC (1987). *Directions in Engineering Research: An Assessment of Opportunities and Needs*. Washington, DC: National Academy Press.

NRC (2012). *Research Universities and the Future of America: Ten Breakthrough Actions Vital to Our Nation's Prosperity and Security*. Washington, DC: National Academy Press.

Rouse, W.B. (1993). *Catalysts for Change: Concepts and Principles for Enabling Innovation*. New York: Wiley.

Rouse, W.B. (1998). *Don't Jump to Solutions: Thirteen Delusions That Undermine Strategic Thinking*. San Francisco, CA: Jossey-Bass.

Rouse, W.B. (2001). *Essential Challenges of Strategic Management*. New York: Wiley.

Rouse, W.B. (Ed.). (2006). *Enterprise Transformation: Understanding and Enabling Fundamental Change*. New York: Wiley.

Rouse, W.B. (2007). *People and Organizations: Explorations of Human-Centered Design*. New York: Wiley.

Royal Society (2011). *Knowledge, Networks, and Nations: Global Scientific Collaboration in the 21st Century*. London: The Royal Society.

Salmi, J. (2011). The road to academic excellence: Lessons of experience. In P.G. Altbach & J. Salmi, Eds., *The Road to Excellence: The Making of World-Class Research Universities* (Chapter 11). Washington, DC: World Bank.

Schoemaker, P.J.H. (1995). Scenario planning: A tool for strategic thinking. Sloan Management Review, Winter, 25–40.

Schwartz, P. (1991). *The Art of the Long View: Planning for the Future in an Uncertain World*. New York: Currency Doubleday.

Snow, C.P. (1965). *The Two Cultures: And a Second Look*. Cambridge, UK: Cambridge University Press.

Sternberg, R.J. (2012). Essay on why some colleges can't change. Inside Higher Ed, April 3.

Szollosi-Janze, M. (2005). Science and social space: Transformation in the Institutions of Wissenschaft from the Wilhelmine Empire to the Weimar Republic. *Minerva*, *43*, 339–360.

Taylor, M.C. (2010). *Crisis on Campus: A Bold Plan for Reforming Our Colleges and Universities*. New York: Knopf.

Toyama, K. (2015). Why technology will never fix education. Chronicle of Higher Education, May 19.

Wildavsky, B. (2010). *The Great Brain Race: How Global Universities Are Reshaping the World*. Princeton, NJ: Princeton University Press.

12

EXPLORING THE FUTURE

In this chapter, the individual models presented in earlier chapters come together to form an integrated computational model that enables pursuing various "what-if" questions associated with the four scenarios elaborated in Chapter 11. Policy implications of the findings are then discussed. Possible extensions of the model are considered.

The pieces discussed in earlier chapters include:

- Chapter 3—organizational elements and structure
- Chapter 4—governance and decision making
- Chapter 5—administrative structure and costs, overhead
- Chapter 6—tuition, endowment, etc.
- Chapter 7—P&T process and decision making
- Chapter 8—enrollment, class sizes, and faculty mix
- Chapter 9—proposals and articles submitted
- Chapter 10—rankings and brand value
- Chapter 11—scenarios and transformation

The conceptual organization of the overall model follows the multilevel architecture in Figure 12.1. This architecture was used to identify the phenomena that were discussed and modeled in earlier chapters. It is important to note however that the

---

Universities as Complex Enterprises: How Academia Works, Why It Works These Ways, and Where the University Enterprise Is Headed, First Edition. William B. Rouse.
© 2016 John Wiley & Sons, Inc. Published 2016 by John Wiley & Sons, Inc.

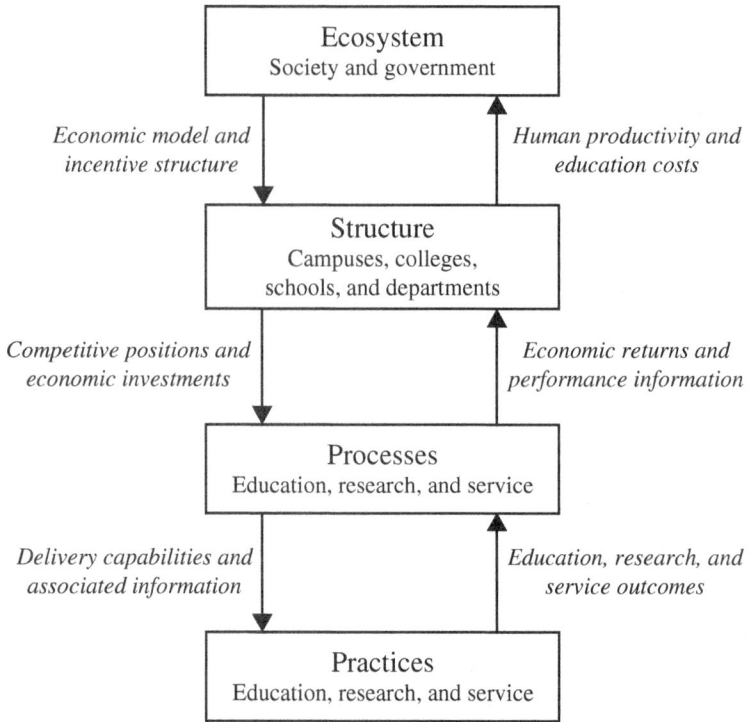

FIGURE 12.1 Multilevel architecture of academic enterprises.

computational framework—a set of linked spreadsheets—does not explicitly employ the structure in Figure 12.1.

The hybrid architecture shown in Figure 12.2 portrays the conceptual impact that universities have on each other. This might suggest that an integrated model would require representation of multiple universities. To avoid such complications, other universities are represented in terms of their aggregate demands for sponsored research and journal publications. These aggregate demands were projected based on data provided by government agencies and journal publishers.

Figure 12.3 portrays the overall flow of variables within the economic model. Not every connection is portrayed, as the figure would become hopelessly messy. Of particular note, students' applications and enrollments are driven by a trade-off between net tuition and brand value. Somewhat simplistically, students seek to matriculate at the highest brand value university that they can afford.

Figure 12.4 shows how all the models come together, with the variables within each model listed. The financial model that follows Figure 12.3 is not shown in Figure 12.4 as it draws revenue and cost data from all the other models. Showing all these linkages would also make this figure quite messy.

The variables listed in Figure 12.4 are elaborated in Table 12.1. The variables included on the dashboard are those that are usually varied to address "what-if" questions.

EXPLORING THE FUTURE 175

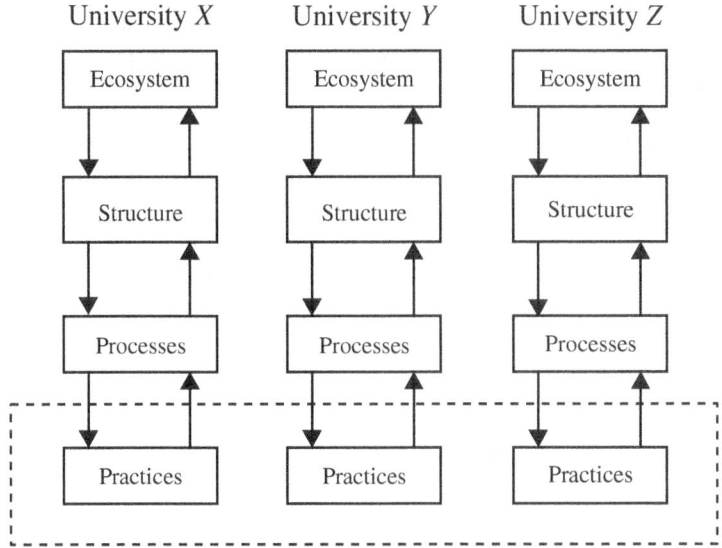

FIGURE 12.2 Hybrid multilevel architecture of academia.

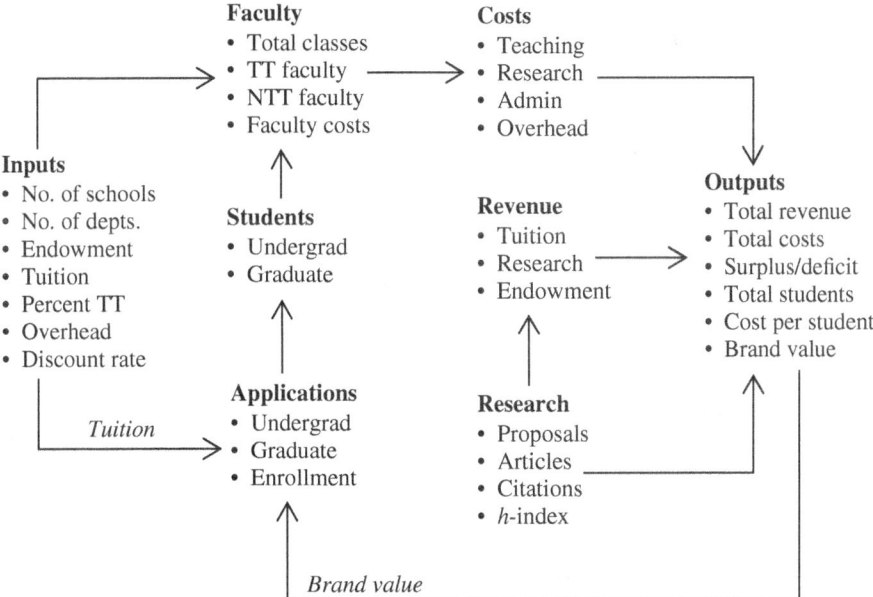

FIGURE 12.3 Overall structure of economic model of academic enterprises.

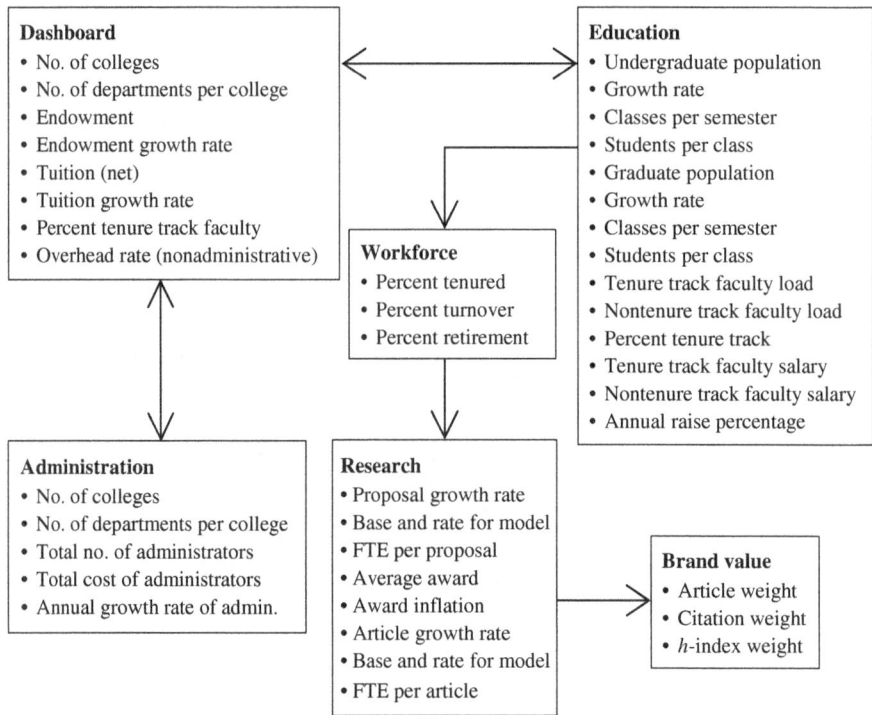

FIGURE 12.4 Computational modules and variables.

The input variables included within individual models can be tailored to particular institutions, but are not usually varied. Of course, any of these variables could be moved to the dashboard. Note that three variables are computed and eight variables are based on empirical data.

SENSITIVITY ANALYSES

Before addressing the four scenarios and how they can be represented in the integrated economic model, it is useful to understand the sensitivity of the overall model's outputs to variations of key variables. Baseline conditions are as follows:

- Colleges = 4
- Departments per college = 5
- Endowment = $150,000,000
- Endowment growth = 5%
- Tuition growth = 3%
- Percent tenure track = 50%

SENSITIVITY ANALYSES

TABLE 12.1 Variables within Overall Model

Dashboard		Base for model	Data
Number of colleges	Input	Rate for model	Data
Number of departments/college	Input	FTE per article	Input
Endowment	Input	**Education**	
Endowment growth rate	Input	Undergraduate population	Input
Tuition (net)	Input	Growth rate	Input
Tuition growth rate	Input	Classes per semester	Input
Percent tenure-track faculty	Input	Students per class	Input
Overhead rate (nonadministrative)	Input	Graduate population	Input
NPV of surplus/deficit	Computed	Growth rate	Input
Finance		Classes per semester	Input
Discount rate	Input	Students per class	Input
Admin		**Faculty**	
Number of colleges	Dashboard	Tenure-track faculty teaching load	Input
Number of departments per college	Dashboard	Nontenure-track faculty teaching load	Input
Total number of administrators	Computed	Percent tenure track	Dashboard
Total cost of administrators	Computed	Tenure-track faculty salary	Input
Annual growth rate of admin	Input	Nontenure-track faculty	Input
Proposals		Annual raise percentage	Input
Proposal growth rate	Data	**Workforce**	
Base for model	Data	Percent tenured	Input
Rate for model	Data	Percent turnover	Input
FTE per proposal	Input	Percent retirement	Input
Average award	Input	**Brand Value**	
Award inflation	Input	Article weight	Input
Articles		Citation weight	Input
Article growth rate	Data	h-index weight	Input

- Overhead rate (nonadministrative) = 50%
- Undergraduate population = 4000
- Graduate population = 4000

Notice that tuition is not included in this list. This is due to tuition being adjusted to cause the net present value (NPV) of the projected surplus or deficit (S/D) to be zero. In other words, for any of the analyses that follow, tuition is set to break even in terms of overall revenue minus costs. This enables rather crisper comparisons of changes of input variables and, later in this chapter, comparisons of scenarios.

Figure 12.5 shows the NPV of S/D for increasing percent tenure-track faculty members for two levels of tuition. NPV accelerates down as this percentage increases, reflecting the increased subsidies of faculty involved in research. Higher tuition, not surprisingly, increases the NPV but does not change the nature

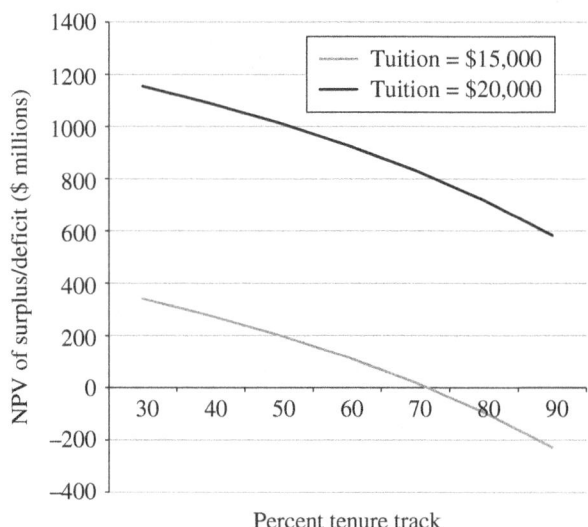

FIGURE 12.5 NPV of surplus/deficit versus percent tenure track.

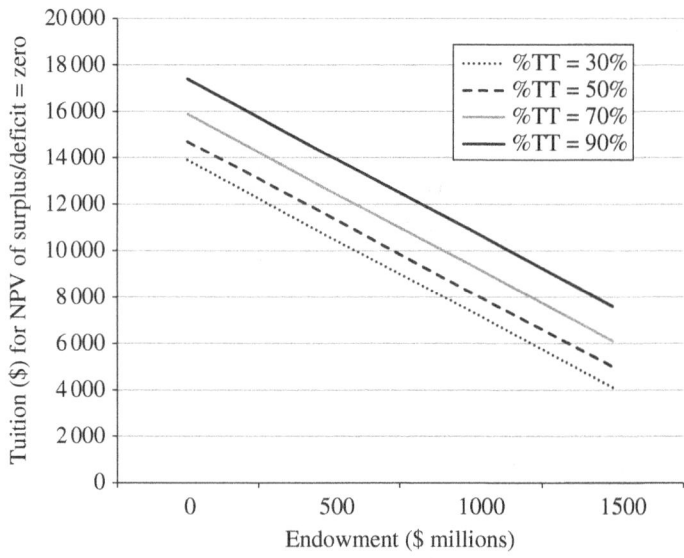

FIGURE 12.6 Tuition for NPV of surplus/deficit = zero versus endowment.

of the phenomenon. The fact that the education function has to subsidize the research function is pretty clear.

Figure 12.6 shows tuition for NPV=0 as a function of levels of endowment and percent tenure track. As one would expect, increasing endowment earnings allows large decreases of tuition. Note that increasing percent tenure track accelerates needs for tuition increases. This nonlinear phenomenon was also evident in Figure 12.5.

SENSITIVITY ANALYSES

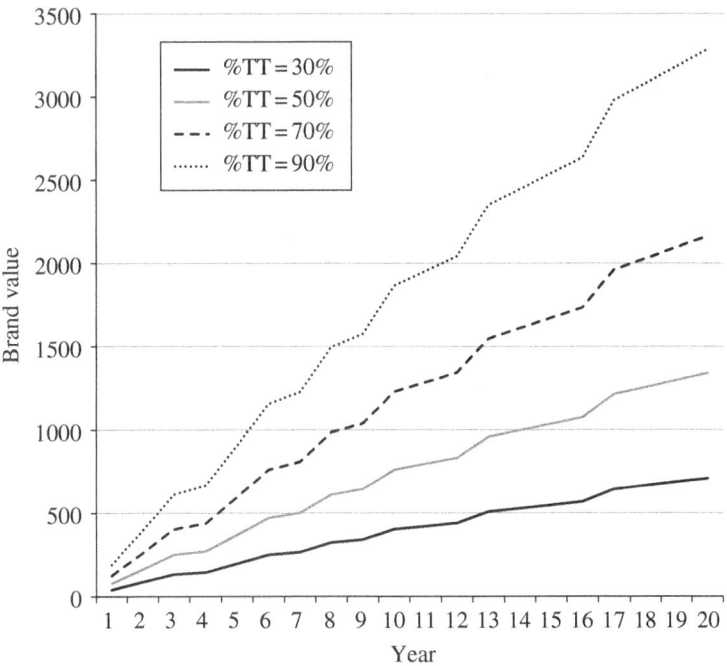

FIGURE 12.7 Growth of brand value for levels of percent tenure track.

Figure 12.7 shows brand value over time. The slope of growth changes with increasing percent tenure track. (Recall that the discontinuities are due to the stepwise nature of the h-index.) Brand value increases with the percentage of faculty members conducting research, publishing articles, and getting cited. Looking at Figures 12.5, 12.6, and 12.7 together, tuition has to be higher at high brand value schools unless they have large endowments

Figure 12.8 projects the number of proposals submitted to NSF and articles submitted to IEEE transactions. The number of proposals to NSF has to accelerate to assure one grant funded every other year. This is due to the 3.6% annual growth of proposals submitted. Number of papers submitted to IEEE decelerates, eventually reaching zero, due to the decreased portion of 50% time that tenure-track faculty members can devote to preparing papers. The acceptance rate of these decreased submissions would further decrease due to the 9% annual growth rate of papers submitted to IEEE.

Figure 12.7 shows steadily increasing brand value. This is due to increasing numbers of tenure-track faculty members. In contrast, Figure 12.8 shows that each faculty member is actually experiencing diminishing returns. Thus, growth is masking underlying productivity deficiencies. The challenge for university leadership is to not let overall metrics delude one into thinking that everything is fine (Rouse, 1998, 2001).

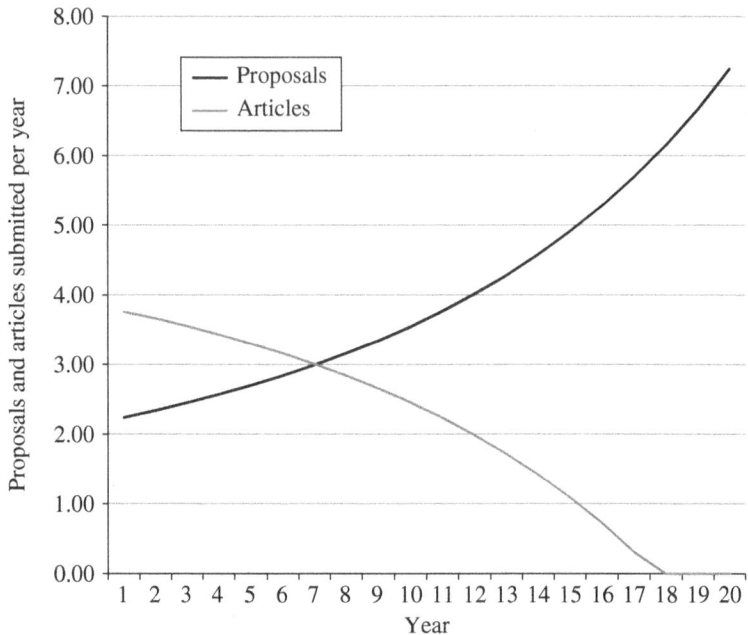

FIGURE 12.8 Proposals and articles submitted per year.

SCENARIO VARIATIONS

Table 12.2 shows a subset of the variables they are affected by the four scenarios from Chapter 11. To project the outcomes for each scenario, I used the same baseline conditions as used for the sensitivity analyses in the previous section. Initial tuition for each scenario is adjusted to achieve NPV = 0. Rather than fixing initial tuition and its rate of growth, I could have adjusted tuition each year to create zero deficits. However, this would make it difficult to compare scenarios.

Clash of Titans

In this scenario of business as usual on steroids, tuition grows steadily by 5% annually. Endowment grows steadily by an aggressive 8%. Percent tenure track is 80% to increase brand value. Percent tenured after the sixth year is 50%. The goal is to retain only the most productive faculty members. The undergraduate population grows slowly at 2%, while the graduate population grows steadily at 6%. Administrative costs grow steadily at 6% as they have in recent years.

Hot, Flat, and Crowded

With competition among global universities intensifying, graduate enrollment decreases by 4% annually reflecting foreign students making different enrollment choices than in the past. Fewer graduate students result in a reduction of tenure-track

TABLE 12.2 Selected Variables versus Scenarios

	Clash of Titans	Hot, Flat, and Crowded	Lifespan Mecca	Network U.
Overall				
Endowment	$150M	$150M	$150M	$150M
Endowment growth rate (%)	8	4	4	2
Tuition (net)	?	?	?	?
Tuition growth rate (%)	5	2	2	1
Percent tenure-track faculty (%)	80	30	30	20
Overhead rate (nonadministrative) (%)	50	50	50	50
Admin				
Annual growth rate of admin (%)	6	3	3	−5
Proposals				
Average award	$100K	$100K	$100K	$100K
Award inflation (%)	3	3	3	3
Teaching				
Undergraduate population	4000	4000	4000	4000
Growth rate (%)	2	2	2	2
Graduate population	4000	4000	4000	4000
Growth rate (%)	6	−4	6	10
Faculty				
Tenure-track faculty load	2	2	2	2
Nontenure-track load	4	4	4	4
Workforce				
Percent tenured	50	30	30	10
Percent turnover	5	10	10	10
Percent retired	10	10	10	10

faculty to 30%. Tuition growth is limited to 2% and endowment growth slows to 4%. Growth of administrative costs is reduced to 3%.

Lifespan Mecca

Enrollment of older students seeking career changes or pursuing retirement interests results in the graduate population growing at 6% per year. The undergraduate population grows more slowly at 2%. Tuition increases are limited to 2% as much of this growth comes from people who are unwilling to pay constantly escalating tuitions. The percent tenure-track faculty decreases to 30% because the MS and perhaps MA degrees being sought require more teaching faculty. Endowment grows slowly at 4%. Growth of administrative costs is limited to 3%.

Network U.

Increased online offerings result in the graduate population growing quickly at 10% annually, while undergraduate population grows more slowly at 2%. Classes become small discussion groups; class sizes vary from traditional numbers to much larger.

The percent tenure-track faculty decreases to 20% as the research enterprise becomes more focused on niches of excellence rather than trying to compete across the board. Tuition growth is necessarily limited to 1% in this highly competitive environment. Endowment grows very slowly at 2%, as most alumni have never set foot on campus. Administrative costs necessarily must decline by 5% annually.

Projections

Figures 12.9, 12.10, and 12.11 show the results of using the variable choices in Table 12.2 as inputs to the overall economic model of the university enterprise. Figure 12.9 portrays student population at year 20 and tuition for NPV equals zero. Note that the tuition does not differ greatly for each scenario. This is due to the model automatically adjusting the number of faculty members to meet demands. This is, of course, easier for nontenure-track faculty members than for those who are tenured.

The student population is depressed for Hot, Flat and Crowded as graduates students choose to enroll at globally equivalent but less expensive universities. Lifespan Mecca attracts older American students that swell the graduate ranks. Not surprisingly, Network U. leads to dramatic growth of online graduate students.

Figure 12.10 portrays the brand value for each scenario. Brand value for Clash of Titans dwarfs the other scenarios, the closet being the baseline. The other three

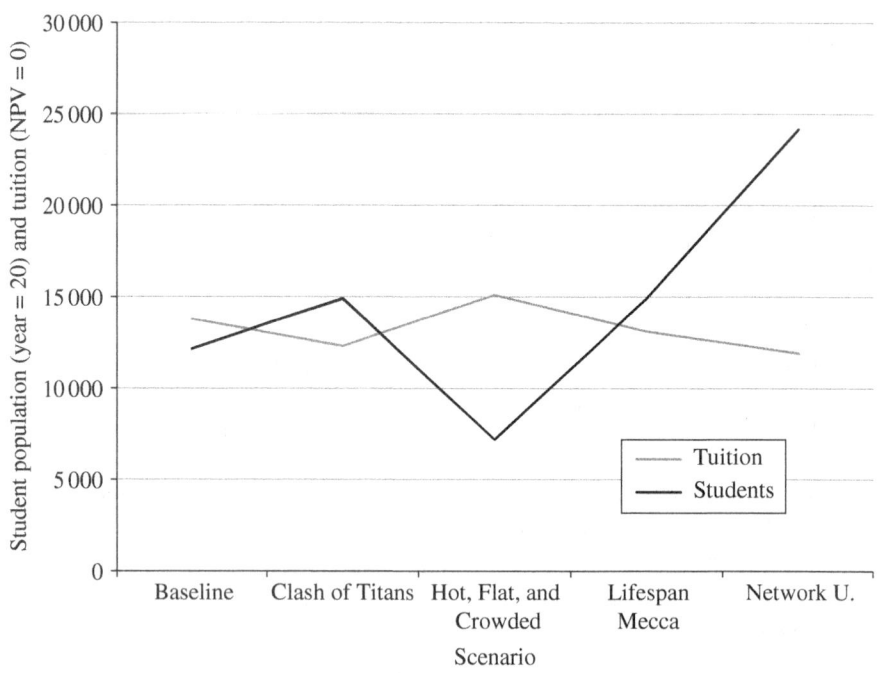

FIGURE 12.9 Student population (year=20) and tuition (NPV=0).

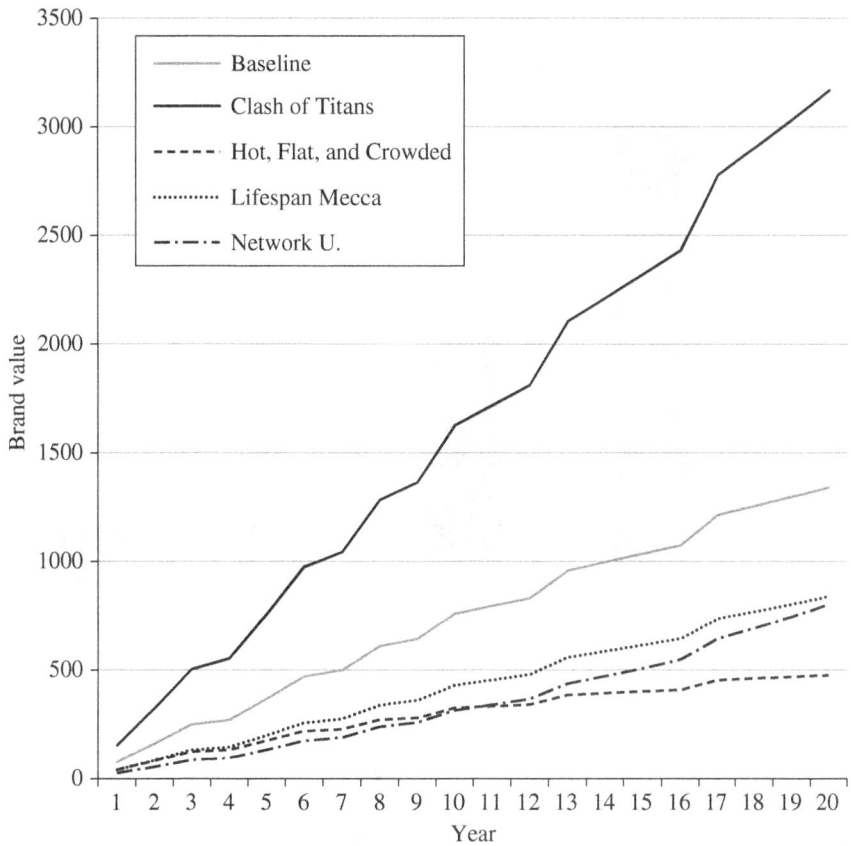

FIGURE 12.10 Brand value for each scenario.

scenarios drive needs to move away from emphases on graduate research conducted on campus. I have found that faculty members often have great difficulty thinking about such alternatives.

We would expect that technology-enabled Network U. to have large classes of remotely connected students, probably very large for lectures and smaller for discussion sections. However, even the discussion classes are likely to be much larger than traditional campus classes. Figure 12.11 shows tuition versus class size in terms of numbers of times larger than the baseline.

The impact is fairly dramatic. As class sizes increase, the overall model automatically reduces numbers of faculty members, which consequently substantially reduces costs. A rapidly growing student body (see Fig. 12.9) while costs of delivery are plummeting, which enables cutting tuition from $12,000 per semester to $2000.

Thus, an undergraduate degree would cost $16,000 in total, assuming it requires eight semesters to earn enough credits to graduate. Of course, by this point the notion

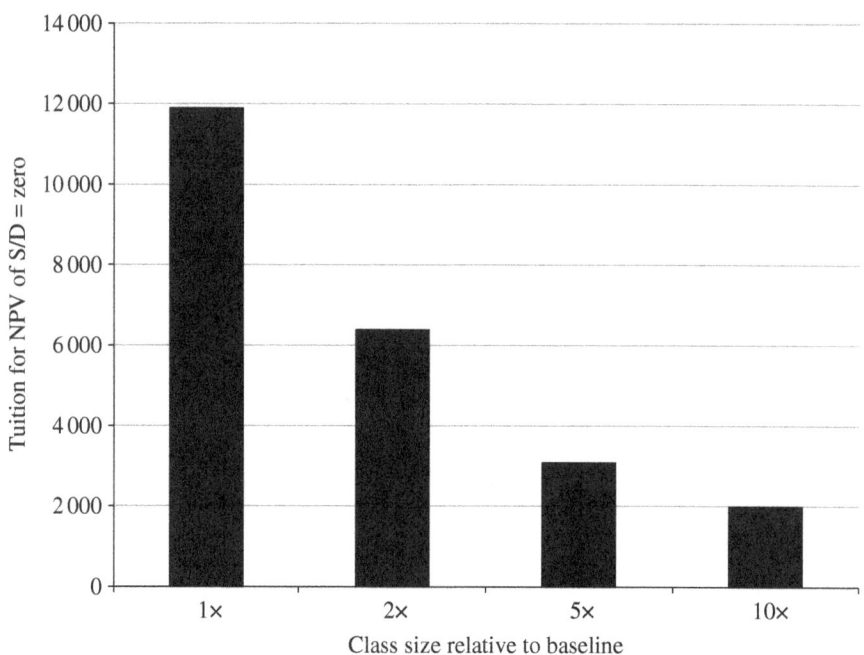

FIGURE 12.11 Tuition (NPV = 0) for class sizes relative to baseline.

of semesters may be completely obsolete. Pricing will probably be by the course. How courses are bundled will be up to each student. Alternatively, pricing might be by the module, with students mixing and matching the modules to gain the knowledge and skills they seek.

This scenario easily causes one to consider what the university should do with its sizable investment in bricks and mortar. One possibility is that this infrastructure mainly serves the resident undergraduate student population, while the graduate population needs limited numbers of traditional classrooms and, of course, no dormitories and dining halls.

An overall comparison of these scenarios is as follows:

- Baseline: Keeps revenues and costs balanced across years with modest brand value, lower than Clash of Titans but higher than the other three scenarios
- Clash of Titans: Begins with slight deficit and then generates growing surplus as student population grows; brand value is strong due to high percent of tenure-track faculty
- Hot, Flat, and Crowded: Leads to declining graduate enrollments and, in later years, steadily increasing deficits; brand value plummets
- Lifespan Mecca: Leads to strong growth of graduate enrollments and essentially zero deficits; brand value increases, relative to Hot, Flat, and Crowded, because of more faculty members being needed to serve increased enrollments

- Network U.: Leads to exploding graduate enrollments; increasing class sizes, enabled by technology, dramatically lowers costs; brand value steadily decreases as larger class sizes lead to reduction of faculty size; initial deficits are replaced in later years by huge surpluses

POLICY IMPLICATIONS

What are the strategic and tactical policy implications of the results found for this set of scenarios? These implications can be considered at two levels—across scenarios and within scenarios. The within-scenario challenges and opportunities are likely to influence how across-scenario issues are best addressed.

Across Scenarios

My experience has been that research universities most readily relate to Clash of Titans. If strong growth of the graduate student population is accompanied with large, yet acceptable, tuition growth as well as endowment growth, the previous results show that pursuit of this scenario is fully viable. Problems will arise, however, if the other scenarios become salient.

The loss of graduate revenues portended by Hot, Flat, and Crowded could be quite difficult to sustain. I recently asked a provost, "What would be the consequences if, for some political or economic reasons, all the Chinese graduate students disappeared?" He responded, "We would lose $35 million in tuition revenues."

I asked, "What is Plan B?" He said, "Well, all research universities would be in this situation." I replied, "That does not sound like a Plan B." To be fair, the university in question is working to diversify its foreign graduate student population, which would decrease the chances of a single point of failure. Better yet would be approaches to encouraging more American graduate students, but this would require investing scarce resources in much larger stipends.

Lifespan Mecca could help this situation substantially, as the results shown previously indicate. Professional educational offerings, supported by students' employers, plus affordable lifelong learning offerings could replace the graduate students lost to Hot, Flat, and Crowded. However, they would not be research students. Nevertheless, developing these offerings would be good hedges against the downside of Hot, Flat, and Crowded.

Network U. could change the whole fabric of the university. As discussed in earlier chapters, guides on the side would replace many or most of the sages on the stage. Flipped classes would become a norm. Online enrollment would soar. Faculty members' interpersonal skills would become core competencies. Students' sense of affiliation would mainly relate to the value being provided by the education. The value of campus amenities and athletics would diminish.

One might embrace this scenario or simply choose to hedge against it. This would, at least, involve investing in capabilities to provide high-value online offerings. One hybrid possibility would be investing in these capabilities to provide the Lifespan

Mecca offerings noted previously. This would provide the competencies, as well as some infrastructure, to enable scaling up when Network U. becomes increasingly prevalent.

The bottom line is that one cannot just choose one of the scenarios. All of them must be addressed if only to define early warning signals of their emergence. More strategically, investments in Lifespan Mecca and Network U. constitute hedges against Hot, Flat, and Crowded. A balanced investment portfolio across all scenarios is likely to be the best approach. It will mean that one cannot put all the eggs in the Clash of Titans basket, as that could be quite risky.

Within Scenarios

Clash of Titans presents a particularly difficult challenge. The current success model at most research universities requires faculty members to work harder and harder to achieve less and less success. Universities need to broaden their views of "gold standard" sponsors beyond NIH and NSF to include other first-rate sponsors such as the National Aeronautics and Space Administration and the Office of Naval Research. Private foundations and industry sponsorship should be increased.

Universities also need to broaden their views of "gold standard" journals beyond current "A" journals. They should emphasize citations rather than impact factors, the irrelevance of which was discussed in Chapter 11. A paper that earns 100+ citations in a low impact factor journal should be seen as a home run, not something to be dismissed.

University presidents, provosts, deans, and promotion and tenure committees need to communicate these changes to their faculties, particular junior faculty members. If everyone continues to pursue the old success model, there will be a lot less success, leading to pervasive frustration of junior faculty and much waste of human and financial resources. The outsourcing of evaluations of junior faculty needs to be tempered by more internal assessments.

Many universities have envisaged keeping Hot, Flat, and Crowded at bay by creating global campuses, the idea being that those who eschew matriculating in America can earn the same credential in Dubai or Singapore. There are merits to this idea but also limits. I have experienced many faculty members of foreign extraction advocating the launch of a new campus in their native country. Campus leadership has encouraged this to the extent that the talent on the home campus was diluted. Having a branch of CMU or MIT in every country is inevitably unsustainable, particularly in terms of brand value and quality of education.

On the other hand, making a Network U. version of CMU and MIT globally accessible makes much more sense. An interesting hybrid involves pursing a year or two online and the rest of the degree on campus. The key is for the university to make the investment to assure high-quality online offerings that lead to the advertised knowledge and skills. This is not simply a matter of putting one's PowerPoint slides on the web. Proactive engagement of students in the learning experience requires that educators design this experience, monitor its evolutions, and constantly improve it.

Lifespan Mecca requires careful attention to what students, ranging from midcareer professionals to eager-to-learn retirees, want and need to gain to achieve their educational aspirations, for example, promotions, new jobs, or simple mastery of history, music, or political science. Many traditional faculty members do not like to teach professionals and see history, music, and political science as "service courses." Success in this arena, therefore, may mean many fewer traditional faculty members.

Summary

Developing a strategy for addressing each scenario is necessary before one thinks through how investments in these strategies might be leveraged across scenarios. The key point is that one does not know what mixture of these scenarios will emerge over the next 10–20 years. One's strategy across these scenarios needs to leverage opportunities while also hedging the downsides associated with these futures (Rouse, 2006).

EXTENSIONS

There is a variety of extensions of the model presented here that could be useful, especially for particular universities who might like to tailor the model to their circumstances. The current model assumes open-loop behavior. More specifically, decision makers stay with their strategy over the years, independent of the outcomes of earlier years. The extensions needed involve adding feedback loops to represent how decision makers react to the evolving state of the enterprise.

A good example of this is setting tuition. The current model allows deficits in any particular year as long as the NPV is positive over the number of years projected. It would be easy to add a feedback loop that zeroes any deficits by increasing tuition. I avoided this to enable easier comparisons of scenarios by assuring that NPV equaled zero across scenarios. That makes sense for this exposition, but might not for an individual university.

Another example is staffing. The model currently only projects expected values of the variables of interest. However, variability in recruiting, turnover, and retirements could result in needs for new faculty members that are greater or lesser than the current model projects. For this feedback loop to be meaningful, variability would have to be added to the model.

Another interesting extension could involve space planning, a major issue at most universities. The model projects growth in terms of more students in more seats in more classrooms. However, the availability of space is not considered. Creation of space is a long lead-time consideration that is subject to varying opportunities and constraints at different universities. Thus, the addition of a space-planning model would need to be tailored to a particular university.

Since gifts and endowments often play a key role in building projects, adding space planning would require diverting a portion of endowment earnings from the general fund where the model now funnels these earnings. This portion could be channeled into a capital budget. However, this would also require the following extension.

The current model operates off an income statement, independent of any balance sheet. Capital investments are not considered. Thus, investments in buildings, laboratories, equipment, etc. are not included in the current financial model. To, in effect, add a balance sheet to the model, it would be necessary to input initial assets, liabilities, and so on. This would only make sense for a particular university.

The current model, indeed the whole book, pays little attention to the humanities and arts. How do the humanities and arts fit into the model of research universities? I have assumed that these faculty members teach a lot and generate much tuition revenue. They may pursue scholarship, but it is not subsidized by release time. This is a convenient but very simplistic assumption.

The scholarly outputs of faculty members in the humanities and arts obviously contribute to a university's brand value. However, this has not been studied as extensively as science and technology. I searched for data on research sponsors and publishing outlets related to humanities and arts. I found various opinion pieces, but almost no data.

Eugene Garfield's creation of citation indices in the 1960s began in science and technology and was later extended to the social sciences. These sources enriched the study of bibliometrics. My guess is that federal funds support such studies because science and technology can be seen as directly linked to creation of intellectual property, subsequent economic activities, and creation of economic growth.

As shown in the following text, NIH and NSF have much larger budgets than the National Endowment for the Humanities (NEH) and the National Endowment for the Arts (NEA):

- NIH: $30.6 billion
- NSF: $7.3 billion
- NEH: $148 million
- NEA: $146 million

Thus, humanities and arts are budgeted less than 1% of what science and technology are budgeted. If I were to include the budgets of other research agencies, the percentage would get much skimpier.

Nevertheless, humanities and arts make enormous contributions to the academic ecosystem. They are key to educating fully rounded students that understand history, politics, aesthetics, ethics, etc. Yet, these elements of the university are not organized and operated like the science and technology elements. Thus, they are difficult to model in the ways elaborated in this book. Without the types of data supporting the models presented here, there would be great risk of making projections totally based on opinion, in my case uninformed opinion.

I felt compelled to limit the range of discourse in this book to evidence-based formulations. To the extent that I have speculated, it has been in terms of parameter variations for "what-if" explorations of the scenarios. The projections associated with these explorations should not be considered predictions. Their purpose was to provide insights into key trade-offs and risks.

CONCLUSIONS

This chapter has brought together all the pieces of the puzzle elaborated in this book. The elements of the education and research enterprise have been integrated to enable projecting the likely consequences of several rather disparate scenarios of the future of the academic enterprise. We cannot predict what mix of these scenarios will actually emerge. However, we can argue that universities need strategies and investments that enable robust responses to whatever mix emerges.

Fundamental change is in the offing. Higher education cannot remain the poster child for runaway costs. We need a healthy, educated, and productive population that is competitive in the global marketplace. If the population is not educated, it will not be healthy. If the population is not productive, it will not be competitive. The pieces all fit together. We have to make it happen.

REFERENCES

Rouse, W.B. (1998). *Don't Jump to Solutions: Thirteen Delusions That Undermine Strategic Thinking*. San Francisco, CA: Jossey-Bass.

Rouse, W.B. (2001). *Essential Challenges of Strategic Management*. New York: Wiley.

Rouse, W.B. (Ed.). (2006). *Enterprise Transformation: Understanding and Enabling Fundamental Change*. New York: Wiley.

INDEX

Abbey, C.W., 147, 151
abuse, 66
academic disciplines, 89, 90, 158
academic ecosystem, 188
academic enterprise, 174, 189
academic spending, 76
academic values, 52
Accenture Innovation Center, 86
Accenture, 38, 127
accreditation, 29
accrediting agencies, 52
accrediting bodies, 105
Acord., S.K., 122, 137
active learning techniques, 92
activity-based cost accounting, 169
adjunct faculty, 3
administration, 11, 51, 57
administrators, costs, 59
administrators, nominal salaries, 62
administrators, number, 59
Advisor Series, *136*
aesthetics, 188
Africa, 4

Age Discrimination Act of 1986, 93
Agency for Healthcare Research and Quality, 81
agents, intelligent, 9
agents, learning, 9
agility, 7, 9
agriculture, 72
AI, 125
Air Canada, 136
Air Force Office of Scientific Research, 81
Air Force Research Laboratory, 81
Air Force, U.S., 36, 136
Alexander, J.H., 146, 151
Altbach, P.G., 27, 155, 169
Amazon, 164
ambiguity, 52
America, 15
American Cancer Society, 126
American Institute for Research, 76
American Recovery and Reinvestment Act, 131
American Society for Engineering Education, 24, 27

Universities as Complex Enterprises: How Academia Works, Why It Works These Ways, and Where the University Enterprise Is Headed, First Edition. William B. Rouse.
© 2016 John Wiley & Sons, Inc. Published 2016 by John Wiley & Sons, Inc.

Anderson, R.D., 27
Andrews, L., 51, 56
Apple II, 157
appointments, 38
Archibald, R.B., 76, 87
architecture, 2
architecture, multilevel, 1
Army Research Laboratory, 81
Army, U.S., 127
articles, accepted, 134
articles, published, 99, 139
articles, submitted, 134
arts, 188
Ashenfelter, O., 93, 103
Asia, 4, 154
astronomy, 167
athletics, 31
athletic spending, 76
Attridge, J.M., 106, 119
audit, 86
Auguste, B.G., 31, 39
Auguste, P., 17
automotive marketplace, 126
autonomy, 155

BankBoston, 31, 39
Banks, J., 167
Barbarossa, F., 16
baseline conditions, 176
Baumol, W.J., 75, 87
Bavorick, H., 156, 170
Beecher, T., 91, 103
behaviors, conflicts, 6
Beijing, 156
Belkin, D., 79, 87
Bell, A.G., 47
benchmarking, 63, 144, 155
benefits, 94, 153, 154
Bennis, W., 45, 55
Berlin, 27
bibliometrics, 121, 122, 188
binomial cumulative distribution, 100
binomial probability density function, 100
biology, 5, 125
biomechanics, 37
Boards of Regents, 51, 54, 84
Boards of Trustees, 51
Bobby Dodd Institute, 72
Bok, D., 25, 27, 150, 151

Bonnycastle, C., 21
book series, 67
Boone, M.E., 5, 13
Bornmann, L., 123, 137
botany, 167
Bowen, H.R., 75, 87
Bradford's law, 123
brand development, 49
brand value, 12, 92, 118, 139, 146, 174, 179, 184, 186
Braun, T., 148, 151
Brazil, 154, 156
Breiding, A., 95, 103
bricks and mortar, 184
bricks-and-mortar strategies, 167
Bridgewater College, 78
Brown, A.B., 146, 151
Brown University, 19
Buchanan, James, 21
budgets, 41, 112
Burka, P., 51, 56
burning platform, 47
Bush, V., 15, 22, 27
business functions, 164
business models, 4, 158
business process improvement, 153, 158
Byron, G.G., 167

calculus, 91
Caldera, L., 158, 170
California Institute of Technology, 22, 23
California State University, 79
Cambridge University, 27, 154
Cambridge University Press, 17
Campos, P.F., 79, 87
campuses, 2, 33
capital budget, 188
Card, D., 93, 103
career awards, 143
Carnegie Institute of Technology, 22
Carnegie Mellon University, 22, 186
Carrell, S.E., 91, 103
cartels, 105
cellphones, 164
Center for Complex Systems and Enterprises, 38, 86, 137
Center for Human-Machine Systems Research, 37
certification, 157

Chameau, J-L., 49
change, 42
change, catalytic mechanisms, 43
change, fundamental, 189
Chappatta, B., 78, 87
Charan, R., 43, 56
chemistry, 125, 167
Chile, 155
China, 155, 156
Christensen, C.M., 4, 13, 29, 39, 154, 158, 170
Chronicle of Higher Education, 58, 93, 103, 106
Chuang, I., 157, 170
Churchill, W., 111
Cisco, 31
citation of articles, 130
citations, 121, 141, 186
cities, 6
city management, 75
Civil War, 168
Clarke, M., 151
Clash of Titans, 160, 161, 165, 168, 180, 182, 184, 185, 186
classes per semester, 114
Clough, W., 49, 143
CNN, 164
Cold War, 156
Coleman, C., 157, 170
Coleridge, S.T., 167
collaboration, 42, 156
collaborative learning, 167
collaborative learning techniques, 92
collaborative networks, 156
colleagues, 78
colleges, 2, 33
colleges, land-grant, 22
colleges, sports, 31
Collins, J.C., 42, 43, 56
Columbia University, 19
Colvin, G., 43, 56
command and control, 7
commitment, sustained, 44
Committee on Committees, 53
committees, 116
communications, 49, 66
Communist Party, 55
competency, 158
competition, 161

competitive advantage, 106, 163
competitive position, 2, 163
competitors, 153
complex adaptive systems, 6, 26
compliance, 66
computational modeling, 5
computational modules, 176
computer science, 154
conflict management, 64
conflict of interest, 66
conflicts, across organizations, 64
conflicts, behaviors, 8
conflicts, goals, 8
control, 6, 9, 57
Control Data Corporation, 157
control theory, 110
controversy, 50
Coordinated Science Laboratory, 35
coordination, 57
Cornell University, 19, 21, 22
corporatization, 58
Cortese, D.A., 7, 13
cost analyses, 76
cost disease, 75, 76, 154
cost efficiencies, 164
costs, 71, 153, 154
costs, administrative, 1
costs of conformity, 67
costs, out-of-control, 1
Cota, A., 31, 39
courses, 109, 113
courses, modules, 54
Covey, S.R., 45, 56
Craig, D.D., 147, 151
creative destruction, 154
creativity, 97
Crecine, J.P., 143
credentials, 108, 166
credit card debt, 108
crisis, 153
CSGNET, 162, 170
Cubism, 168
culture, 42
culture of conformity, 97
cultures, university, 162
Curis Meditor, 137
curricula, 41, 51, 109
curriculum committee, 109
Cynefin Framework, 5

Daly, C.J., 94, 103
Davy, H., 167
Day, G.S., 43, 56
deans, 57, 186
de Boer, F., 31, 39
decision theory, 110, 111
decision threshold, 100, 103
decomposition, 5
Dee, J.R., 94, 103
Defense Advanced Research Projects Agency, 23, 81, 136
defense economics, 72
degree programs, 105, 108
degrees, 41
Delft University of Technology, 30, 36, 52, 125, 136
delivery of education, 110
Dell, 164
Delta Cost Project, 76, 87
demands, aggregate, 174
DeMillo, R.A., 4, 13, 29, 40, 111, 119, 157, 168 170
demographic trends, 161
department chairs, 57
Department of Energy, 81
Department of Homeland Security, 81
departments, 2, 33
dependencies, legislative, 80
design of experiments, 110
design thinking, 110
Designer's Associate, 136
Designing Your Life, 110
Desrochers, D.M., 31, 40, 76, 77, 87
de Weck, O., 128, 137
Dewey, J., 168
digital natives, 161, 163
diminishing returns, 149, 179
direction, 57
disagreement, 52
disciplines, 2, 140
discounted cash flow, 5
dogma, 168
Dollar General, 127
Dow Jones Industrial Average, 142
Dubai, 186
Duderstadt, J.J., 25, 27
Dutch language, 36

Eaton, A., 20
economic growth, 188
economic model, 2, 71, 85
economic returns, 2
economics, 5, 71
economics, higher education, 1 73
ecosystem, 1, 32
ecosystem, society and government, 32
Edison, T., 47
editors, 124, 137
education, 34, 35, 68, 72
education, affordable, 158
education, campus-based, 154
education, online, 154
education, paradigms, 110
education, programs, 12, 105
effectiveness, 156
efficiency, 7, 26, 156
Efron, L., 51, 56
Einstein, A., 168
Eisenberg, E.M., 51, 56
Eliot, T.S., 168
emergence, 5
Emory University, 126
employer-employee relationship, 55
enabling technologies, 166
endowed chairs, 81, 82
endowments, 178, 179
engineering economics, 72
Engineering Enterprise, 64, 67
Engineering Experiment Station, 30
engineering science, 24
England, 16
English language, 36
Enlightenment, 20
Enterprise Support Systems, 136
enterprise, 2
enterprise, architecture, 5
enterprise, as-is, 42
enterprise, system, 4
enterprise, to-be, 42
enterprise, transformation, 1, 42, 137, 153, 158, 160, 169
enterprises as systems, 37
ethics, 188
Europe, 4, 156
evaluation, 57
evidence-based decision making, 85, 150
evidence-based formulations, 188

INDEX 195

excellence, 155
executive programs, 71
expectations, 165
experiences, fundraising, 82
experiences, governance, 52
experiences, leadership, 49
experiences, research, 125
experiences, spin-off, 135
external advisory boards, 52
Eyring, H.J., 4, 13, 29, 39

Facebook, 160
facilities, 41, 161
faculty, 51
faculty, governance, 53
faculty, impact, 91
faculty, morale, 94
faculty, nature, 90
faculty, positions, availability, 92
faculty, retirement, 93, 115
faculty, roles, 90
faculty, turnover, 93, 115
faculty, worklife, 94
Fahey, L., 161, 170
false acceptance, 99
false rejection, 99
Federal Aviation Administration, 81
Federal student loan program, 71, 74
Fedoruk, A., 122, 138
feedback loops, 187
fee for service, 7
Feldman, D.H., 76, 87
Feynman, R., 166
Figlio, D., 92, 103
financial reports, 84
financial resources, 51
Financial Times World University Rankings, 5
Fitzsimmons, J.M., 148, 151
Fleischer, V., 83, 88
flipping the classroom, 111, 185
F1 student visa, 71
food stamps, 92
forces, driving, 160
forces for change, 153
France, 16
Fredrickson, C., 92, 103
freedom, 94
Freeman, J., 159, 170

Freud, S., 168
Friedman, T.L., 160, 162, 170
fundamental change, 4, 47
fundamental limits, sports, 36
funding, 121, 124, 155
fundraising experiences, 82
future, 13

Gandhi, 168
Garcia, D., 144, 151
Gardner, H.E., 168, 170
Garfield, E., 188
Gatorade, 135
Genentech, 135
General Electric, 142
General Motors, 127
genomics, 48
George Mason University, 39
George, W., 42, 56
Georgia Institute of Technology, 12, 22, 30, 37, 49, 50, 53, 61, 66, 84, 86, 95, 97, 110, 111, 113, 127, 136, 137, 141, 146, 154, 162
Georgia Tech Business Network, 64
Georgia Tech Research Institute, 31
German, language, 10
Germany, 15, 19, 167
Ghaffarzadegan, N., 93, 104
Gibb, K., 122, 138
GI Bill, 80
Ginsberg, B., 58, 69
Gladwell, M., 7, 13, 142, 146, 151
globalization, 91, 106, 153, 154, 161
global marketplace, 189
global parity, 162
Global Research Benchmarking System, 147
global supply chain management, 164
goals, 51, 125
goals, conflicts, 6
gold standard, 186
Goldwater-Nichols Act, 80, 135
Gosnell obsolescence, 123
governance, 11, 41, 51, 155
governance, experiences, 52
governance, government-led, 18
governance, models, 15
governance, student-led, 18
governing boards, 51

government, 32
government sponsors, 127
grade point average, 143
grades, 166
graduate curriculum committees, 54
Graduate Record Exam, 114
Graham, M., 168
grants funded, 99
Great Recession, 93, 112
Greeks, 16
growth of brand value, 149
guides on the side, 111, 167, 185
guilds, 90

Hannan, M.T., 159, 170
Hansen, J., 157, 170
Harantova, M., 146, 151
Harley, D., 122, 137
Harris, M., 107, 119
Harvard University, 4, 19, 25, 83, 154
HarvardX, 157
Hatch Act, 30, 80
Haughwout, A., 78, 88
healthcare delivery, 1, 34, 72, 75, 126
hedging the downsides, 187
Herschel, W., 167
heterarchy, 7, 9
hierarchy, 7, 9
higher education, 1
higher education, bubble, 74
higher education, costs, 75
h-index, 130, 139, 147, 148
hiring, 41, 51
Historically Black Colleges and Universities, 21
history, 5, 10, 111, 187, 188
Hitler, A., 111
Ho, A.D., 157, 170
Holmes, O.W., 167
Holmes, R., 167, 170
H1-B visa, 106
Hong Kong, 155
Horn, M.B., 158, 170
Hot, Flat, and Crowded, 160, 162, 165, 180, 182, 184, 185,186
Howard, D.J., 124, 137
Howard Hughes Medical Institute, 124
Hoyt, J.E., 146, 151
human capital economics, 72

human capital theory, 73
human-centered design, 136
human cognition, 125
humanities, 91, 188
human-machine systems, 110
Humboldt University, 20

IBM, 127, 164
IBM PC, 157
IEEE Transactions, 128, 129, 149, 179
Immersion Lab, 38, 86, 126
impact factors, 95, 186
impedances, 29
improvement, 26
incentives, 7, 26
incentive structures, 2
independent agents, 8
India, 155, 156
Indiana University, 126
indirect costs, 77
industry sponsors, 127
inertia, 145
influence, 7
information and control requirements, 125
inhibitions, 7
innovation, 97, 105
innovation, technology driven, 106
insights, 156, 189
Institute of Medicine, 126
institutional supports, 91
integrity, 168
intellectual property, 12, 121, 135, 188
interactive visualization, 126
Internet, 157
investments, 2
Ioannidis, J.P.A., 123, 137
IP, *see* intellectual property
IPAMM, 131, 137
iPhone, 160
Italy, 16
IT system, 113
IT systems, legacy, 55
Ivy League, 19

James, W., 167
Japan, 156
Jarvis, J., 61
Jayaram, K., 31, 39
Jefferson, T., 20, 21

jobs, 121, 162
jobs, high-paying, 106
jobs, prospects, 107
Jobs, S., 47
Johns Hopkins University, 59
Johnsrud, L.K., 93, 104
John Wiley & Sons, 67, 136
joint appointment, 36
Joint Services Electronics Program, 35, 127
journal, articles, 3
Journal of Systems Engineering, 128
journals, A., 96
journals, impact factors, 148
journals, role, 123
judgment, 50

Kamenetz, A., 163, 170
Keats, J., 167
Kelic, A., 106, 119
Kershaw, I., 111, 119
Kezar, A., 91, 92, 104
Khan, S., 157
Khurana, R., 158, 170
Killian, J.R., Jr., 23, 27, 80, 88
Kindle, 160
King Charles I, 17
King Henry III, 18
King Henry II, 17
Kirshstein, R.J., 76, 77, 87
know-how, 121
knowing-doing gap, 43
knowledge, 156, 184
Koenig, H.F., 146, 151
Korea, 155
Kotter, J.P., 42, 56
Kouzes, J.M., 45, 46, 56
K-12 education, 33, 106, 163

Laboissiere, M.C.A., 31, 39
Laboratory for Measurement and Control, 36
labor economics, 72, 73
labs, 161
Laird, F.N., 124, 137
land-grant colleges, 22
laptop, 107
Lariviere, R., 75, 83, 88
Larson, R.C., 93, 104
Latin America, 4

Lazerson, M., 31, 40
leaders, 41
leadership, 7, 9, 11, 41, 42, 57, 94, 150
leadership, change agent, 47
leadership, experiences, 49
leadership, time, 44
learning, 6
learning experience, 186
learning outcomes, 105
Ledford, H., 77, 88
Lee, D., 78, 88
lessons learned, 83
Levin, R.C., 24, 28
Lewis, M., 168, 170
libraries and networks, 125
lifespan education, 163
Lifespan Mecca, 160, 162, 166, 181, 182, 184, 185, 186, 187
limits of modeling, 125
Lincoln, A., 21, 108
loans, government-backed, 29
Lockheed Martin, 38, 127, 136, 165
Lombardi, J.V., 4, 13, 90, 104, 147, 151
Lopez, G., 157, 170
Lotka's law, 123
loyalty, 94

Mabe, M., 123, 138
Maccoby, M., 45, 56
MA degrees, 163
Malaysia, 155
management, 7
managerialism, 91
managers, 41
manufacturing, 164
Manufacturing Extension Partnership, 31
MA offerings, 166
Marcus, J., 59, 69
Marginson, S., 141, 151
Marine Safety International, 135
market forces, 26, 153
marketing, 49, 66
market perceptions, 164
Martin, R.E., 78, 88
Massachusetts Institute of Technology, 21, 22, 23, 31, 80, 125, 135, 146, 154, 186
massification, 91
mastery, 158
Maxey, D., 91, 92, 104

MBA, 158
McDonald, M., 78, 87
McGurn, W., 74, 88
Medicaid, 71, 92
Menand, L., 158, 170
mental models, 37
mentor-mentee relationship, 55
merit, 155
metrics, 148
Mexico, 155
Miller, A.I., 170
Mills College, 78
Mills, N., 58, 69
mission, 10, 29
mission, creep, 75
mission, critical, 48
mission, statement, 30
Mitchell, J., 78, 88
MITx, 157
model dashboard, 85
model extensions, 173, 187
modeling, 110
modules, 184
money, 11, 65, 71
MOOCs, 4, 154
morale, 116
Morrill, J.S., 21
Morrill Land-Grant Acts, 10, 15, 20, 80
Morrill Land-Grant Act, Second, 80
Morse, P.M., 125, 138
Motorola, 112, 164
MS degrees, 163
MS offerings, 166
multiattribute utility model, 111
Murphy, A., 51, 56
music, 187
Mussolini, B., 111
Mutz, R., 123, 137

nanoscience, 48
Narin, F., 122, 138
National Academies, 50, 106, 119, 126, 147, 156, 170
National Academy of Engineering, 126
National Academy of Medicine, 126
National Aeronautics and Space Administration, 23, 81, 127, 186
National Endowment for the Arts, 188

National Endowment for the Humanities, 188
National Institutes of Health, 23, 48 81, 89, 96, 98, 99, 124, 127, 131, 132, 138, 159, 186, 188
National Institute of Standards and Technology, 31
National Research Council, 24, 28, 140, 151, 156, 171
National Science Board, 131, 138
National Science Foundation, 15, 23, 25, 48, 81, 89, 92, 96, 98, 99, 104, 122, 124, 127, 131, 133, 143, 159, 179, 186, 188
National Science Foundation Act, 23
Nature, 123, 128, 129, 138, 141, 148, 149, 151
Naval Research Laboratory, 81
Navy, U.S., 127
needs-beliefs-perceptions model, 159
Nelson, R.R., 72, 88
net present value, 177
Network U, 160, 163, 166, 181, 182, 183, 185, 186
Newell Rubbermaid, 165
niche dominance, 4
Nigeria, 155
Nobel Prize, 25, 50, 113, 141, 147
Noel-Levitz, 146, 151
Nokia, 112
nonlinear dynamic behavior, 8
nontenure-track faculty members, 90, 114
normal science, 2
North Carolina State University, 67, 83
Northcutt, C., 157, 170
Northern Light, 38, 127
North, G., 163, 171
NRC, *see* National Research Council
NSF, *see* National Science Foundation
number of articles published, 147
number of citations received, 139, 147

Oakland Athletics, 168
objectives, 57
O'Connell, A., 110, 119
Office of Naval Research, 81, 186
Ohio State University, 22
Olympics, 1996, 143
O'Meara, K., 95, 103
online networks, 158

INDEX

open-access records, 122
operations research, 112
opportunity, 153
organizational change, 42, 153, 158, 159
organizational culture, 94
organizational delusions, 160
organizational design, 7
organizational momentum, 94
organizational structure, 15
organizing, 57
Osakwe, A., 122, 138
Osuji, J., 122, 138
O'Toole, J., 45, 55
outlets, valued, 3
outsourcing, 169
overall model, 132
overall model, variables, 177
overall organizations, 164
overhead, 48, 84, 86, 169
overview, 10
Oxford University, 154
Oxford University Press, 17, 28, 33

package delivery company, 164
P&T, *see* promotion and tenure
paradigms, 96, 97
participatory leadership, 94
patents, 121
Paulsen, M.B., 73, 88
payment for health outcomes, 7
peer quality, 73
peer review, 52, 96, 121, 122
Pennsylvania State University, 22
pensions, 47
people conflicts, 65
percent tenure track, 178
perceptions, 165
Perez-Pena, R., 135, 138
performance, 153
performance, evaluation, 61
performance, management, 4
Perry, R., 51
Petersen, R., 157, 170
Pfeffer, J., 43, 56
Ph.D. programs, 124, 143
Phelps, E.S., 72, 88
philanthropic gifts, 81
Phillips, E.D., 147, 151
physics, 125

Picasso, P., 168
Pierce, C.S., 167
Pilot's Associate, 136
Piscopia, Elena Lucrezia Cornaro, 18
planning, 42, 51, 57
plans, 125
PLATO, 157
Poincaré, A.H., 168
Poli, R., 5, 13
policy decisions, 41
policy flight simulator, 126
policy implications, 173, 185
political science, 187
politics, 188
Pope Gregory IX, 17
Porras, J.L., 42, 56
portable smart devices, 157
Posner, B.Z., 45, 46, 56
postdocs, 124
postdoctoral appointments, 95
power, 7
practices, 1, 35
presidents, 57, 186
President's Advisory Committee on Science, 24, 28
prevention and wellness, 126
prices, 153
Princeton University, 19
prisons, 71
Pritchard, A., 122, 138
private good, 33, 73
probabilities of success, 3
processes, 1, 34
product offerings, 164
Product Planning Advisor, 136
productivity, 72, 133
productivity, deficiencies, 179
productivity, paradox, 150
professional education offerings, 166
professor of the practice, 54
profit, 121
programs, 47
programs, doctoral, 109
programs, master's, 109
programs, undergraduate, 109
projections, 182
promotion, 41, 51, 89
promotion and tenure, 2, 11, 34, 89, 162
promotion and tenure, committees, 186

promotion and tenure, letters, 96
proposals, 3, 61
proposals submitted, 134
provosts, 57, 186
publications, 61, 89
public endowment, 75
public good, 33, 73
Public Health Service, 81
public policy, 5
public sector, 66

Qualcomm, 112
quality, 153, 156
quality, management, 4
queuing networks, 110
Quiggin, J., 142, 151

radios, 164
Randall, R.M., 161, 170
rankings, 12, 139
rankings, determinants, 143
rankings, schemes, 140
registrar, 113
Reich, J., 157, 170
relativity, 168
Rensselaer Polytechnic Institute, 20, 28
Rensselaer, S., 20
research, 12, 34, 35, 68, 121
research centers, 35
research centers, leading, 48
research, experiences, 125
research funding, 84, 89
research funding, agencies, 81
research infrastructure, 37
research, model, 128
resource allocation, 41
revenue, 71, 78, 121
revenue, fundraising, 81
revenue, government dependencies, 80
revenue theory, 76
revenue, tuition, 79
risks, 189
Robinson, K., 110, 119
Rogers, W.B., 21
Romans, 16
Roosevelt, F.D., 111
Rosser, V.J., 93, 94, 104
Royal Society, 156, 171

Ruhvargers, A., 142, 151
rules of the game, 34, 97
Russia, 155

Sabatier, G., 106, 119
safety, 72
Sage missile defense system, 23
sages on the stage, 111, 167, 185
salary, 94
salary structure, 47
Salmi, J., 155, 169, 171
Samuelson, P., 166
Sanburn, J., 156, 170
Sanchez, A., 106, 119
Saracevic, T., 123, 138
Scally, J., 78, 88
scam, 108
scenario, development, 161
scenario, high growth, 116
scenario, low growth, 116
scenario, variations, 180
scenarios, 160, 189
scenarios, transformation, 12
Schapiro, M., 92, 103
Schoemaker, P.J.H., 161, 171
Scholastic Achievement Tests, 143
School of Industrial and Systems Engineering, 144
schools, 2, 33
Schouten, J.W., 146, 151
Schumpeter, 31, 40, 80, 88
Schwartz, P., 161, 171
science, 91, 123, 141
Science Citation Index, 123
Scientific Atlanta, 30
scripts, 125
Search Aeronautics, 136
search committee, 51
Search Technology, Inc., 136
Seglen, P.O., 148, 151
selectivity, 113
self-organization, 6, 7, 9
semesters, 113, 184
Senge, P.M., 46, 56
senior design, 112
sensitivity analysis, 176
Serban, N., 7, 13, 75, 88, 125, 138
services, 34, 35, 68, 99, 153, 164
services, activities, 116

INDEX

services, offerings, 164
Shanghai Jiao Tong University, 140, 141
Shanghai Jiao Tong University Academic Ranking of World Universities, 5
Shelley, P.B., 167
short courses, 136
Shribman, D., 108, 119
Silver, A., 156, 170
simulation-oriented computer-based instruction, 135, 157
Singapore, 155, 186
Situation Assessment Advisor, 136
Skevington, J.H., 148, 151
skills, 156, 184
Smart Grid, 111
smartphone, 107
Smith, B., 29, 40
Smith-Lever Act, 30, 80
Snow, C.P., 91, 104, 167, 171
Snowden, D.J., 5, 13
Soares, L., 158, 170
social networking, 42
social sciences, 5
Social Security, 72
social technology, 163, 167
society, 32
Soter, K., 92, 103
space, 11, 65, 71, 86
space, hoarding, 86
spin-off businesses, 121
spin-off experiences, 135
sponsored research, 143
sponsors, valued, 3
staffing, 187
staffing patterns, 77
stakeholders, 165
Stalin, J., 111
Stanford University, 23, 77, 83, 84, 110, 142
Starbucks, 164
state legislatures, 51
State of Georgia, 72
State of Oregon, 83
State of Rhode Island, 74
STEM challenges, 106
Sternberg, R.J., 159, 171
Stevens, 84, 97, 98
Stevens Institute of Technology, 22, 30, 38, 63, 66, 83, 95, 110, 113, 126, 137

Stevens Institute Series on Complex Systems and Enterprises, 67
stewards of the status quo, 46, 154
stimulus funds, 112
stipends, 109, 124, 185
stock offering, 121
Strategic Planning Advisor, 136
strategy, 42, 51, 166
Stravinsky, I., 168
structural inertia, 159
structure, 1, 10, 29, 33
structure, heterarchical, 2
structure, hierarchical, 2
student engagement, 142, 186
student loan debt, 1, 108
student population, 106
student ratings, 91
students per class, 114
student trade-off, 74
submission of articles, 128
submission of proposals, 131
subsidies, federal, 80
subsidies, tuition, 80
success model, 186
Sullivan, L.H., 34, 40
summer salaries, 124
sunshine laws, 51
supertanker propulsion system, 135
sustainable energy, 34
Sutton, R.I., 43, 56
Svoboda, P., 146, 151
Sweet Briar College, 78
system dynamics model, 106
systems, chaotic, 5
systems, complex, 5
systems, complicated, 5
systems, dynamic, 6
systems engineering, 130
Systems Engineering Research Center, 86
systems, nonlinear, 6
systems, simple, 5
Szollosi-Janze, M., 167, 171

tablet, 107
talent, 146, 155, 161
Taylor, M.C., 158, 171
teacher ratings, 61, 96, 99
teaching, 89
teaching experiences, 112

teaching faculty, 3
teaching loads, 114
teams, pickup, 52
teamwork, 42
technological diffusion, 72
technological innovation, 156
technology, 153, 157
Technology Investment Advisor, 136
Tennenbaum Institute, 38, 82, 86, 137, 160
Tennenbaum, M., 49
tenure, 41, 47, 51, 89, 115, 158
tenure, decision making model, 99
tenure, decisions, 95
tenure, packages, 98
tenure-track faculty, 3, 90, 114
tenure-track positions, 90
Texas A&M University, 51
Tharoor, I., 156, 170
Thayer, S., 20
The Economist, 103, 154, 155, 170
The Netherlands, 36
Thomson-Reuters, 123, 138
threat, 153
time pressure, 94
Times Higher Education, 130, 138, 140, 141
tipping points, 7, 146
Title IX Amendment, 80
Tojo, H., 111
Toutkoushian, R.K., 73, 88
Toyama, K., 157, 171
trade-offs, 189
training, 72
transcript, 108
transformation, 29
transformation, framework, 160, 164
transformation, scenarios, 153
tribes, 91
Trowler, P.R., 91, 103
Tsinghua University, 55, 156
tsunami of talent, 156
Tufts New England Medical Center, 126
Tufts University, 30, 95, 112, 125, 126
tuition, 71, 78, 174, 183, 187

Umbach, P.D., 92, 104
United Airlines, 37
United Auto Workers, 55

United Nations University International Institute for Software Technology, 147
United States, 4, 10, 15, 106, 156
United Technologies, 127
universities, evolution, 15
University of Arizona, 22
University of Berlin, 19
University of Bologna, 16, 27, 33
University of California, Berkeley, 23, 54, 83
University of California, Los Angeles, 54
University of California, Merced, 54
University of California, San Diego, 54
University of California, San Francisco, 135
University of Cambridge, 18
University of Connecticut, 55, 98, 127
University of Florida, 22, 135
University of Illinois, Urbana-Champaign, 22, 23, 30, 35, 39, 50, 53, 83, 95, 97, 110, 113, 127, 125, 135, 157
University of Maryland, 22, 95
University of Michigan, 25
University of Oregon, 75
University of Padua, 17, 28
University of Paris, 17, 28, 33
University of Pennsylvania, 19, 126
University of Rhode Island, 55, 74
University of Rochester, 59
University of Texas, 51, 83
University of Virginia, 21, 28
University of Wisconsin, 22
university organization, 60
UPS, 164
urban ocean systems, 126
urban resilience, 34
US Atomic Energy Commission, 23
US Department of Education, 105
US Military Academy, 20
US News & World Report, 5, 58, 67, 121, 140, 151, 161
US patents, 25
Uzoka, F.-M, 122, 138

value of education, 72, 107
value proposition, 1, 29, 105
values and norms, 41
Vanderbilt University, 59, 126
van der Klaauw, W., 78, 88
Van der Werf, M., 106, 119

Van Noorden, R., 148, 151
Victoria's Secret, 164
vision, 41, 42, 150
visualization, 110
von Humboldt, A., 20
von Humboldt, W., 10, 15, 19

Waffle House, 66
Wall Street Journal, 105, 119
Wal-Mart, 164
Ware, M., 123, 138
Wawrzynski, M.R., 92, 104
Web of Science, 123
weightings, 140
Weimar Republic, 167
Wellcome Trust, 124
Wells, B.H., 106, 119
Werner, R., 123, 138
West, J.E., 91, 103
West Point, 20, 28
what if questions, 5, 173, 174
Whirlwind, 23

Whitehill, J., 157, 170
Wildavsky, B., 154, 171
Wilhelmine Empire, 167
Williams, J.J., 157, 170
Winston, G.C., 73, 88
Wood, P.W., 74, 78, 88
Worcester Polytechnic Institute, 22
work activities, 164
workforce model, 105, 114
work physiology, 37
World Bank, 155
world-class status, 150
World War I, 168
World War II, 10, 15, 22

Xue, Y., 93, 104

Yale University, 19, 24, 83

Zagonel, A., 106, 119
Zipf's law, 123
Zappula, F., 128, 138